David J. Owen, MLS, PhD

The Herbal Internet Companion
Herbs and Herbal Medicine Online

More pre-publication
REVIEWS, COMMENTARIES, EVALUATIONS . . .

"**A**ll Internet/computer-literate persons with a serious interest in herbs and herbal medicine will warmly welcome Dr. Owen's exciting new book. This reading audience will include practicing herbalists and teachers of botanical medicine and alternative medicine in general, as well as librarians working with significant holdings in herbal medicine in particular, or with substantial holdings in alternative medicine in general. And, of course, this audience naturally will include many other students of herbs and herbalism. This book is also especially timely and badly needed, for there is at present no other single, current source that even comes close to providing so much electronic research help on these botanical subjects. Dr. Owen's book also has a wealth of related, more practical information, such as where to find those herbal chat rooms and elusive botanical bulletin boards. Although the focus is certainly on Web research of herbs and herbal medicine, this book should also be of some use for electronic research in alternative medicine in general, or naturopathy. Nor is the information presented here strictly limited to American interests and concerns, for Web sites done in (or about) other English-speaking countries are also well represented. *The Herbal Internet Companion* is an exceptionally well-balanced and practical guidebook for all serious herbal researchers as well as others with related interests."

D. Bryan Stansfield, PhD, MLS
Library Director,
Southwest College
of Naturopathic Medicine
and Health Sciences,
Tempe, Arizona

The Haworth Information Press®
The Haworth Herbal Press®
Imprints of The Haworth Press, Inc.
New York • London • Oxford

The Herbal Internet Companion
Herbs and Herbal Medicine Online

The Herbal Internet Companion
Herbs and Herbal Medicine Online

David J. Owen, MLS, PhD

The Haworth Information Press®
The Haworth Herbal Press®
Imprints of The Haworth Press, Inc.
New York • London • Oxford

Published by

The Haworth Information Press® and The Haworth Herbal Press®, imprints of The Haworth Press, Inc., 10 Alice Street, Binghamton, NY 13904-1580.

Cover design by Marylouise E. Doyle.

Library of Congress Cataloging-in-Publication Data

Owen, David J.
    The herbal internet companion : herbs and herbal medicine online / David J. Owen.
      p. cm.
    Includes bibliographical references and index.
    ISBN 0-7890-1051-8 (alk. paper)—ISBN 0-7890-1052-6 (alk. paper)
    1. Herbs—Therapeutic use—Computer network resources—Directories. 2. Internet. 3. Web sites—Directories. I. Title.

RM666.H33 O975 2002
025.06'615321—dc21

2001039562

To my parents, and my sister Christine,
in gratitude for their love and support;
to Warren Kennell and David Dickson for their friendship;
to Miss Collins for her wisdom, guidance, and encouragement;
and to the loving memory of my sister Veronica.

ABOUT THE AUTHOR

David J. Owen, PhD, is Education Coordinator/Librarian for the Basic Sciences in the Library and Center for Knowledge Management at the University of California, San Francisco. He also holds an appointment as Assistant Clinical Professor in the UCSF School of Pharmacy.

Dr. Owen holds a bachelor's degree in the biological sciences, a PhD in microbiology, and a master's degree in library and information science. Before moving to UCSF, he worked as an information specialist for biotechnology companies. He has published in both scientific and library journals.

CONTENTS

Preface

Nature, whose sweet rains fall on unjust and just alike, will have clefts in the rocks where I may hide, and secret valleys in whose silence I may weep undisturbed. She will hang the night with stars so that I may walk abroad in the darkness without stumbling, and send the wind over my footprints so that none may track me to my hurt: she will cleanse me in great waters, and with bitter herbs make me whole.

Oscar Wilde, *De Profundis*

One of the most remarkable developments in recent years has been the reemergence in the United States of medical therapies and procedures commonly referred to as "alternative medicine." To the consternation of many health care providers, medical practices that were once widely regarded as relics of more ignorant times are resurfacing to claim a place in mainstream American medicine. Acupuncture, homeopathy, traditional Chinese medicine, and massage therapy are no longer confined to the fringes of medical care but are competing with mainstream medicine for the attention of the public. One of the areas gaining considerable publicity, and at the same time generating a lot of controversy, is herbal medicine. It sometimes seems that not a day goes by without the appearance of yet another newspaper article or television news item about St. John's wort or *Ginkgo biloba*. Once confined largely to health food stores, herbal preparations are now prominently displayed on the shelves of modern pharmacies and can be readily purchased via the Internet. They are now widely used by the general public to treat a variety of conditions, from depression to sexual dysfunction, often without the consent or knowledge of a primary physician.

I think it is correct to say that it was only in the mid-1990s that physicians in the United States began to appreciate just how popular herbal products had become. In my current position as a health sciences librarian at University of California, San Francisco (UCSF), my duties include guiding health care practitioners to the most reliable drug information resources. A health science library fields a large number of drug-related questions from both professionals and the public, and the standard pharmaceutical literature is vast and complex. Fortunately, a large body of up-to-date and reliable information resources is available, and an experienced user can readily access accurate information on prescription and over-the-counter drugs.

When I first assumed my position at the UCSF health sciences library almost ten years ago, I knew that *Martindale: The Extra Pharmacopeia* had some basic information on herbs, and I usually referred patrons to this reference text.[1] I also remember that on one or two occasions, out of desperation, I sent someone to the history of medicine collection to consult early twentieth-century publications, such as *The Dispensatory of the United States of America*.[2] However, such questions were few and far between. Then, around 1997 to 1998, I began to notice a significant increase in the number and range of these questions. Interestingly, more and more of these queries were coming from nurses and physicians. The standard U.S. drug reference literature was often woefully inadequate, though I thankfully discovered Varro Tyler's two wonderful books, *The Honest Herbal* and its companion volume, *Herbs of Choice*.[3,4] The past few years have witnessed a remarkable increase in the number and quality of herbal resources available, not least of which is the publication of an English-language translation of Germany's seminal Commission E monographs.[5] However, locating reliable information about herbs can still pose something of a challenge for both the librarian and the health care provider.

Many health care workers are justifiably concerned that the whole area of alternative medicine is replete with unreliable, exaggerated, or misleading information. Unfortunately, mainstream health care providers still consider only a few resources authoritative, so health sciences librarians have a major problem trying to build a core collection in herbal medicine. The 1999 Brandon-Hill list of core medical journals and textbooks for the small medical library, generally acknowledged to be a good source for the most current and authoritative medical texts, lists only three textbooks and one key journal for the entire alternative medicine area.[6] Much of the information for alternative medicine is still to be found in the so-called "gray literature," such as trade journals, pamphlets, conference proceedings, and market research reports; thus, it is often difficult to identify and obtain.[7]

Many people are now seeking health information on the Internet. Information that used to be available only in research journals or in the collection of a local medical library is now being published on Web sites that can be accessed by anyone with a home computer and an Internet Service Provider (ISP). Though much of this health-related information is being provided by credentialed health professionals and organizations, many Web sites are simply glorified advertisements or are run by nonprofessionals with questionable or no qualifications. Type in a search term such as *herb* or *herbal* in any Internet search engine and you retrieve thousands of "hits," most of them for commercial sites selling herbal products. Physicians, nurses, and pharmacists are justifiably concerned about the potentially harmful effects resulting from consumers and health professionals using this information inappropriately.

Beginning in 1998, in response to requests for herbal information from faculty and students here at UCSF, I began to examine how herbal information is presented on the Internet, compiling a list of those World Wide Web sites which I considered to be the most valuable. This formed the basis for a short paper I wrote for a library journal, listing and describing these sites.[8] One impetus for compiling such a list was the apparent paucity of such guides in the literature, and my belief that the few published sources, even those which had begun to appear in mainstream medical journals, were far from complete and usually overlooked some of the better sites. Since 1998, my initial list of Web sites has grown considerably longer, and major developments in the United States and in Europe have significantly influenced the availability of online information about herbs.

When I first began looking at herbal resources, most of my attention was focused on finding information for those herbs most widely used in the United States and Europe, such as St. John's wort, echinacea, and valerian (see Table 1), and during the writing of this book, I maintained that initial focus. However, many mainstream physicians are investigating herbs used in Chinese medicine or in other traditional healing systems that are, as yet, not very widely known in North America. So, as the book grew, along with the number of Web sites, I found myself seeking information on more esoteric herbs, such as devil's claw *(Harpagophytum procumbens)* from Africa or cat's claw *(Uncaria tomentosa)* from Peru.

TABLE 1. Top Ten Best-Selling Herbs in the United States in 1998-1999[1]

---

1. Ginkgo *(Ginkgo biloba)*
2. St. John's wort *(Hypericum perforatum)*
3. Ginseng *(Panax ginseng, P. quinquefolius, Eleutherococcus senticosus)*
4. Garlic *(Allium sativum)*
5. Echinacea (*Echinacea* spp.)/Goldenseal *(Hydrastic canadensis)*[2]
6. Saw Palmetto *(Serenoa repens)*
7. Kava-Kava *(Piper methysticum)*
8. Pycnogenol [3]/Grape Seed
9. Cranberry *(Vaccinium macrocarpum)*
10. Valerian *(Valeriana officinalis)*

Source: Blumenthal, M. Market Report. *HerbalGram,* Number 47, pp. 64-65, 1999.

[1]Ranked by sales.
[2]These two herbs are commonly combined with each other.
[3]This is a patented extract derived from the bark of pine trees.

    This book is intended to serve as more than just an annotated bibliography of herbal Internet sites. My hope is that this book will be a guide to using the Internet for research into all aspects of herbal medicine. For convenience, included Web sites are grouped into chapters based on the primary type of information to be found on them, such as regulations and standards, or side effects, adverse reactions, and drug interactions. Each chapter begins with introductory information that hopefully places the Web sites in context, providing the reader with some background knowledge necessary for a better understanding of the current political and social issues surrounding herbal medicine in North America and Europe.

    In general, I have not thought it necessary to include detailed directions for navigating many of these Web sites, nor for locating specific posted resources. The design and organization of Web sites, as well as their corresponding navigational tools, have undergone considerable advances since the early days of the World Wide Web. After all, extensive collections of resources on particular subjects are not much use if they are hard to locate or navigate. Many sites now include a Web site map, which is a visual listing or representation of its contents. Others provide an integrated search engine to help users access information: just typing in one or more terms is usually sufficient to find relevant documents.

This book is divided into sixteen chapters, dealing with different aspects of herbal medicine such as botanical information, historical research, clinical evidence for efficacy, adverse reactions, regulatory issues, and consumer information. Each chapter begins with a short introductory section, followed by an alphabetical listing of selected Web sites or resources providing access to information in this area, along with the URL, description, and guide to the type of information to be found. Obviously, some sites provide information in several areas so they may be listed in more than one chapter. Since many health professionals are now concerned about the reliability of health information found on the Internet, the opening chapter addresses this issue. A glossary of terms commonly encountered in the current herbal literature is also included. While on the subject of terminology, you will notice that I frequently use the term Complementary and Alternative Medicine, or CAM, when referring to the whole area of "alternative medicine." The reason for this is explained in Chapter 14.

Please note that any information you find on these Web sites is not meant to substitute for the advice provided by a physician or other health care professional. Nor are there any intentional endorsements of any partical product, or treatment recommendations of particular diseases or health-related conditions.

David Owen

Chapter 1

Herbal Medicine and the Internet

Be careful of reading health books. You may die of a misprint.

Mark Twain

HERBAL INFORMATION RESOURCES

In the preface to the 1981 edition of the *The Honest Herbal,* Dr. Varro Tyler, a respected authority on herbal medicines, opined:

> More misinformation regarding the efficacy of herbs is currently being placed before consumers than at any previous time, including the turn-of-the-century heyday of patent medicines.[1]

In the years since this was written, major developments, both in the United States and in European countries, have considerably improved the amount of reliable information on herbs that is now available to both the general public and health care providers. However, although both the number and quality of information resources have improved, the increased consumer demand for herbal preparations has resulted in a proliferation of both print and online publications that are of variable quality and reliability. The newcomer to herbal medicine often has to navigate through a maze of unsubstantiated claims and anecdotal information, a large percentage of it written by nonprofessionals or herbal product manufacturers. Difficulties associated with finding substantiated data are often compounded by what seems to be the strong mystical or New Age slant of much herbal writing—something that you will not find in *The Physicians' Desk Reference* (PDR). In addition, and perhaps most important, to those trained in the methodology of scientific medicine, traditional herbal literature seems to be antiscientific, as well as antagonistic to mainstream medicine and its public health watchdogs, such as the Food and Drug Administration (FDA).

HERBAL INFORMATION AND THE INTERNET

Problems associated with locating reliable herbal information have been exacerbated by the arrival of the Internet and the associated proliferation of health-related Web sites. Personal computers are now ubiquitous, as well as easy to use, and sophisticated health care consumers are beginning to use the Internet to educate themselves about their own health. A vast amount of health care data is available on the Internet. The current estimate is that over 10,000 sites now provide some type of health information.[2] According to a report published by the Pew Research Center in November 2000, 52 million American adults have now used the Web to get health or medical information, and 47 percent of those who sought health information for themselves say the material affected their decisions about treatments and care.[3]

Many health professionals and organizations are justifiably concerned about the quality of this Web-based health information because of the extreme variability in its quality and the effect of commercial interests on its content.[4] It is technically very easy to publish on the Web and can be done by practically anyone with a computer, inexpensive software, and access to the Internet. Unlike such documents as peer-reviewed journal articles, Web-based information resources face little regulation or standardization: their quality is often determined solely by the organizations and individuals who publish them. This is of particular concern because misinformation or the incorrect use of medical information can be harmful. As the public interest in herbs grows, so does the number of Web sites, with a large amount of unsubstantiated reports and misinformation. Many unique Web-based sources are posted by individuals or organizations that operate outside mainstream medicine. A 1997 study of Internet herbal information by students at Albany College of Pharmacy compared claims made about eleven popular herbal products with data from peer-reviewed journals: they concluded that only 45 percent of associated claims were true, 6 percent were false, and 2 percent were meaningless.[4]

Educating health professionals and consumers to assess the quality of health-related information found on Web sites has now become something of a priority. This particularly applies to herbal medicine and other areas of alternative medicine because of the current lack of established authoritative resources. Furthermore, it should be remembered that the World Wide Web is only one section of the Internet; access to much valuable data on the use of herbs exists also in the more informal milieus of electronic mailing lists or newsgroup postings, where information is often anecdotal and difficult to evaluate.

Several groups, such as Silberg and colleagues, have devised lists of basic key criteria that need to be considered when assessing the quality of a medical Web site.[3] A 1999 article by Kim and collegues provides a useful overview of several such systems used to evaluate and rank Web-based information.[5] During the writing of this book, I consulted several sources providing lists of criteria for the evaluation of Web-based information and, where possible, used these as a guide to the selection of sites for this book (see Table 1.1). However, I must emphasize that in many cases I applied these criteria very liberally. For example, I tried to find background information on the authors of a couple of widely used herbal Web sites. Despite detailed searches and repeated attempts to contact them by e-mail, I still have not been able to adequately identify these individuals. Anonymity associated with Web content is usually highly suspect. However, these sites have been included because they are widely used as information channels by U.S. herbal practitioners.

HOW "STABLE" ARE HERBAL WEB SITES?

It is a given assumption that Web sites are continually changing, daily and even hourly. However, currently it is not quite as chaotic as it may seem. In the two years I spent writing this book, I constantly monitored sites for stability. During this time, some sites reorganized, changed names, or became inactive for a short period of time; however, to my surprise (and relief), only one or two sites "disappeared" during this period, and one of these reappeared in another guise a few months later. Sometimes a site was reorganized and resources were moved and could not be easily traced; in these cases, I easily found the new location by typing a few keywords into one of

TABLE 1.1. Criteria for the Selection of Herbal Internet Sites

- **Source:** What type of organization created or sponsors the site? Sites where the URL ends in *.gov* (government resources) and *.edu* (organization connected to education in some way) may be considered to be the most unbiased and reliable sources. The next reliable type of organization is identified by *.org,* generally indicating a nonprofit agency, though this is often an organization advocating a particular issue or viewpoint, and the information presented may therefore be skewed in favor of this viewpoint. Commercial sites are identified by *.com* and their primary function is to make money.

- **Accuracy:** Is the information accurate? Does the information check out against that found in reliable print sources such as Tyler's *Herbs of Choice, Review of Natural Products,* and journal articles in the mainstream medical literature?

- **Referenced information:** Does the site use authoritative sources to support the information presented? Is the information referenced or linked to other sources? Does it cite research articles in the mainstream literature or well-known and respected resources such as the *Commission E Monographs?*

- **Content:** Are contributions written by respected herbalists or medical researchers? If not, does an editorial board review the information before it is posted? Is there a form of peer review?

- **Currency:** When was the site created? How current is the information? How frequently is the information updated?

- **Depth:** What is the depth of coverage? How detailed is the information?

- **Uniqueness:** How unique is the information? Can it be found more easily elsewhere?

- **Bias:** How balanced is the information? Is it presented in a fair and unbiased manner? Is there a hidden agenda? Does the site promote one particular viewpoint? Is the site sponsored by commercial interests? Is the site from a more "traditional" herbal organization or institution that has a unique philosophical approach to the use of herbal remedies?

- **Design:** Is the information well-organized and "professional" looking? Is the site easy to navigate? Is it easy to find information?

- **Audience:** Is the intended audience clearly stated, e.g., patient or health care provider?

- **Stability:** How long has the site been available? Is it always accessible? Are there ever any technical problems using it?

- **Reputation:** Has the site been recommended or favorably reviewed elsewhere?

- **Interactivity:** Is there a way to contact the organization or author? If e-mail is an option, does the organization or author reply to messages?

the many Internet search engines available. (Go to Danny Sullivan's Search Engine Watch at <http://searchenginewatch.com/> and his Web Searching Tips section for comprehensive information on how to use search engines.)

The following Web sites and documents provide more detailed information and valuable discussions about using evaluation criteria for Web content.

WEB SITES FOR EVALUATING INTERNET MATERIAL

Criteria for Assessing the Quality of Health Information on the Internet (Policy Paper)
  <http://hitiweb.mitretek.org/docs/policy.html>
    (direct link to paper)
  <http://hitiweb.mitretek.org/> (Mitretek home page)

Mitretek, Inc., is a nonprofit company seeking to promote accessible, affordable, quality health care by conducting and supporting appropriate research. Along with representatives of the general public, health care providers, and medical librarians, its Health Information Technology Institute (HITI) has developed a set of criteria for use in assessing the quality of health information on the Internet. These criteria are a useful aid in evaluating whether information is usable and credible. HITI participants include representatives from the American Medical Association (AMA), the Federal Trade Commission (FTC) (see entry for FTC in Chapter 11), the American Pharmaceutical Association (APhA), and the National Center for Complementary and Alternative Medicine (NCCAM) (see entry for NCCAM in Chapter 6).

HITI criteria can be briefly summarized as follows:

1. *Credibility:* Is the information from a well-known and respected source?
2. *Content:* Is the information accurate?
3. *Disclosure:* What is the mission or purpose of the site?
4. *Links:* Do these sites have links to other high-quality sites?
5. *Design:* Is the information easy to find?
6. *Interactivity:* Can you send questions to authors or the Web administrator?

7. *Caveats:* Even though the information is valuable, is the primary function of the site to market a product?

Health On the Net Foundation (HON)
<http://www.hon.ch/>

Based in Geneva, Switzerland, HON describes itself as a nonprofit international foundation with a mission to "promote the effective and reliable use of the new technologies for telemedicine in health care around the world." Part of that mission includes guiding both lay users and medical professionals to reliable sources of health care information on the Internet. HON has developed a Hon Code of Conduct (HONcode) in an effort to help the general public identify high-quality Web sites. The HONcode defines a set of rules designed to encourage Web site developers to follow ethical standards in the presentation of information and to help users know its source and purpose. A site can apply the HONcode seal of approval, which is displayed on the Web site's pages.

Information Quality WWW Virtual Library Web Site
<http://www.vuw.ac.nz/~agsmith/evaln/evaln.htm>

This metasite is a collection of links to a variety of sites and documents that deal with assessing the quality of information resources, particularly those found on the Internet.

Chapter 2

An Herbal Renaissance

The thing that hath been, it is that which shall be; and that which is done is that which shall be done: and there is no new thing under the sun.

Ecclesiastes 1:9
The King James Bible

HERBALISM IN THE UNITED STATES

Herbal remedies were an important component of American medicine right up until the early years of the twentieth century. Medical historians believe that upon their arrival in North America in the sixteenth century, early European explorers began trading information about herbal remedies with the Native Americans they encountered. According to *Medicinal Plants of Native North America,* the various indigenous tribes had over 18,000 uses for herbs.[1] Several of these herbs, such as seneca snakeroot (*Polygala senega* L.), wild cherry (*Prunus virginiana* L.), and balm of Gilead (*Populus candicans* Ait.), actually found their way into some editions of the *United States Pharmacopeia* (USP) or the *National Formulary* (NF). Some herbs sold today, such as echinacea, were originally used by Native Americans for the same purposes they are used for today.

During the nineteenth century, American medical movements, such as American eclecticism, introduced a range of herbs into common use, including purple coneflower (*Echinacea* spp.), goldenseal *(Hydrastis canadensis),* and black cohosh *(Cimicifuga racemosa).*[2] In the early part of the twentieth century, approximately 170 herbal drugs were still listed in the USP and NF.[3] Preparations containing many of today's most popular herbs were being manufactured by

7

large pharmaceutical companies, and in pharmacy schools, pharma-cognosy (the study of the properties of drugs of natural origin) was an important part of the curricula. Some herbal materials, such as the dried roots and rhizome of echinacea, remained in the *National Formulary* right up to the 1950s.[3]

Several factors contributed to the general disappearance of herbal medicines from pharmacy shelves. The end of the nineteenth century saw the rise of scientific medicine and the growth of new disciplines, such as experimental physiology and physiological chemistry. The germ theory of disease revolutionized the treatment of infectious diseases. Major advances in organic chemistry led to the introduction of synthetic chemical drugs, such as the sulfonamide antimicrobials, and the rise of the modern pharmaceutical industry. Pharmacologically based chemical therapeutics with a biochemical emphasis began to displace other types of healing systems. In 1910, Abraham Flexner published his famous report *Medical Education in the United States and Canada,* which recommended that the quality of medical education should be improved by integrating scientific principles and techniques into treatment methods and medical school curricula. The Flexner Report had enormous influence and resulted in the closing of many medical schools that did not subscribe to the biochemical theory of medical treatment.[4] Prior to Flexner, approximately 150 medical schools existed in the United States and Canada, but his report was sufficiently damning that more than half of those schools, which he considered "trade schools," closed. Alternative systems of healing, such as homeopathic medicine and eclecticism, disappeared from center stage.

After the Flexner Report, and with the rise in the belief that pharmaceutical drugs could eliminate all disease, the use of herbal remedies experienced a dramatic decline. Plant extracts did not disappear completely from the American materia medica, however, for around one-quarter of all conventional drugs on pharmacy shelves are still derived one way or another from plants—aspirin, for example, was originally extracted from the bark of willow trees (*Salix* spp.); a heart drug called atropine, from deadly nightshade *(Atropa belladonna);* and the anticancer drug paclitaxel, from the Pacific yew tree *(Taxus brevifolia).*[5]

Up until the mid-1990s, herbal preparations such as valerian and echinacea continued to be available to consumers, but they were largely confined to health food stores. It seemed that with the advent of computers and molecular biology, scientific medicine in Western industrialized societies was about to enter a new Golden Age. Many American health care providers were therefore rather startled when surveys published in 1990 and 1997, by Eisenberg and his colleagues, reported that 42.1 percent of all American adults had used at least one form of alternative therapy in 1997, up from 30 percent in 1990, and that four out of ten Americans had used alternative medicine therapies in 1997, including herbs, megavitamins, massage, self-help groups, and homeopathy.[6,7] Most astonishing was the revelation that the total number of visits to alternative medicine practitioners exceeded visits to all U.S. primary care physicians. Additional surveys, such as that conducted by *Prevention* magazine in 1997, indicated that 32 percent of U.S. adults were using herbal remedies.[8]

The explosion of interest in herbs and other areas of Complementary and Alternative Medicine (CAM) continues unabated. Many practitioners of conventional medicine have become more familiar with CAM therapies or are trying to learn more about them: a 1997 issue of *Nature Medicine* reported that 80 percent of contemporary U.S. medical students now want more training in CAM.[9] There is a burgeoning herbal market, and more than 300 companies now market herbal remedies, including such pharmaceutical giants as American Home Products (AHP) and Bayer Corporation.[10] Herbal preparations are being sold not only in health food stores but in supermarkets, in pharmacies, and on the Internet.

The Web sites and resources in this chapter include online catalogs for prominent libraries with important herbal book and periodical collections as well as bibliographic databases focusing on the history of medicine. Also included are Web sites providing the full text of classic herbal monographs that are still regularly referred to by traditional herbalists. Many classic herbals are a rich source of information that cannot be readily found elsewhere.

WEB SITE RESOURCES FOR HERBAL TEXTS

A Brief History of Herbalism
<http://www.med.virginia.edu/hs-library/historical/herb/intro. html>

This document is part of an online exhibit from the Health Sciences Library of the University of Virginia highlighting four herbals that are milestones in the history of Western herbal medicine: "The Vienna Dioscorides" (*De Materia Medica,* 512 A.D.); *The Badianus Manuscript* (an Aztec herbal of 1552); John Gerard's *Herball, or Generall Historie of Plants* (1597); and Nicholas Culpeper's *English Physician* (1652). The introductory section provides a brief history of herbalism, from the Chinese, through Greek and Roman medicine, to the European Middle Ages.

The British Library
<http://portico.bl.uk/>

The British Library began as the library of the British Museum, dating back to 1753. It separated from the British Museum in 1972. Universally considered to be one of the world's leading national libraries, it provides a wide range of services, including bibliographic, document supply, and information services, as well as exhibitions, publications, and events. Its collection contains over 150 million items, representing every conceivable topic, including herbal medicine. The British Library's online library catalog OPAC 97 lists materials held in several of the library's major collections. To see the range of services offered online, select What's Available Without A Visit.

The British Library's Health Care Information Service produces the Allied and Complementary Medicine Database (AMED). This resource covers a selection of journals in complementary medicine, physiotherapy, occupational therapy, rehabilitation, podiatry, palliative care, and other professions allied to medicine (for more information about AMED, see Chapter 13).

Culpeper's Complete Herbal

According to the *Oxford English Dictionary,* an herbal is "A book containing the names and descriptions of herbs, or of plants in general, with their properties and virtues; a treatise on plants." Nicholas Culpeper's *The English Physician,* published in 1652, the first popular European herbal, was written to make medical knowledge accessible to the lay public. The full title is *The English Physitian: or An Astrologo-Physical Discourse of the Vulgar Herbs of this Nation.* It is probably the most famous herbal and was reprinted as late as 1820. Though much of Culpeper's information was factual, the book contained many references to the planets and astrological signs, or astrology, Culpeper's other main interest. Culpeper is still widely used by traditional herbalists, who refer to his formulations and preparations.

Electronic versions of the complete text, or selected sections, are available on the Internet from the following sources:

Bibliomania.com Ltd
<http://www.bibliomania.com/2/1/66/113/frameset.html>

This is the full text of a nineteenth-century reprint provided by Bibliomania.com Ltd, an online literature library with an extensive library of reference books, biographies, classic nonfiction, and religious texts.

The Cushing/Whitney Medical Library, Yale University
<http://www.med.yale.edu/library/historical/culpeper/
culpeper.htm>

This version from Yale was prepared by Richard Siderits, MD, and colleagues from a copy at the Historical Library, Cushing/ Whitney Medical Library, Yale University.

University of Michigan Digital Library Production Service
<http://www.hti.umich.edu/images/wantz/cul1.htm>

Selected images are available from the University of Michigan's Digital Library Production Service (DLPS).

German National Library of Medicine (Deutsche Zentral-
bibliothek für Medizin)
<http://www.zbmed.de/>

In Germany, medicinal herbs are closely integrated into allopathic medicine. A great deal of herbal information is found in the German pharmaceutical literature. The German National Library of Medicine (ZBMed) is financed by the federal government and serves as the main library for medicine, public health, and associated basic sciences. ZBMed acquires and indexes both German and foreign biomedical literature, concentrating on important journals in biomedical fields as well as on monographic literature. All languages are represented, but special attention is given to German and English publications. Go to the English-language section to search the catalog. The library has a large collection of books on herbal medicine.

Herbals & Early Gardening Books
<http://www.goldcanyon.com/Patten/index.html>

This online catalog is for the Doris and Marc Patten Collection of herbals and early gardening books, located at the Hayden Library's Special Collections Department at Arizona State University. It includes important herbals and gardening works dating from the fifteenth century, including herbals from Western Europe, India, and Latin America, and works describing North American Indian medicines. Each entry is fully annotated and includes selected online images from the publication.

Herbs and Herb Gardening: An Annotated Bibliography and Resource Guide
<http://www.nal.usda.gov/afsic/AFSIC_pubs/srb9606.htm>

This is an extensive annotated bibliography on herbs and herb resources compiled by Suzanne DeMuth at the National Agricultural Library (NAL)'s Alternative Farming Systems Information Center (AFSIC), part of the U.S. Department of Agriculture (USDA). Although focusing on herb gardens and cultivation methods, it includes much useful information on a wide range of books and other publications relevant to the history of the medicinal use of herbs.

HISTLINE (HISTory of medicine onLINE)
<http://igm.nlm.nih.gov/>

The HISTLINE bibliographic database, produced by the National Library of Medicine (NLM), is an index to the journal literature dealing with the history of medicine and its related fields. This valuable resource for researching herbal medicine includes literature about the history of health-related professions, sciences, specialties, individuals, institutions, drugs, and diseases in all parts of the world and from all historic periods. It is updated weekly and has around 190,000 citations. HISTLINE indexes U.S. and foreign publications, including monographs, journal articles, and individual chapters in the published proceedings of symposia, congresses, and the like (see Figure 2.1 for a sample record).

As of this writing, this database is accessible through the NLM's Internet Grateful Med (IGM) system. Future plans include the incorporation of journal records into the PubMed system (see PubMed entry in Chapter 13) and the eventual replacement of IGM by a new NLM Gateway **<http://gateway.nlm.nih.gov/gw/Cmd>**.

Karolinska Medical Library Information Sources in the History of Biomedicine **<http://www.mic.ki.se/History.html>**

The Karolinska Institutet, in Stockholm, Sweden, is one of the world's most prestigious medical research centers. The institute's library provides consumers, health care professionals, and researchers with a large collection of links for diseases, disorders, and related-topic resources. Information Sources in the History of Biomedicine is a large collection of links to Internet resources on the history of medicine.

King's American Dispensatory **<http://metalab.unc.edu/herbmed/eclectic/kings/main.html>**

One of the most important medical movements in nineteenth-century America was inspired by the so-called "eclectics." Started by Wooster Beach, under the name Reformed Medicine, eclecticism rejected adherence to any one theory of therapeutics, and, as the name suggests, believed healing methods should be chosen from a variety

FIGURE 2.1. A Sample Record from the HISTLINE Database

TITLE:	From rudbeckia to echinacea: the emergence of the purple cone flower in modern therapeutics.
AUTHORS:	Flannery MA
AUTHOR AFFILIATION:	Lloyd Library & Museum, Cincinnati, OH.
SOURCE:	Pharm Hist. 1999;41(2):52-9.
SECONDARY SOURCE ID:	HMD/99604119
MAIN MESH HEADINGS:	Drug Therapy/*HISTORY Medicine, Herbal/*HISTORY *Plants, Medicinal

of different systems. At one time, eight eclectic medical colleges existed in this country, the most famous being the Eclectic Medical Institute in Cincinnati, Ohio (see this chapter's entry for Lloyd Library and Museum). Eclectic practitioners were medical doctors with a strong belief in the use of herbal medicines. One of eclecticism's basic tenets was an emphasis on treating a group or pattern of symptoms, usually with small doses of a so-called "specific" remedy. Many eclectic physicians were also scientific herbalists who studied herbs, analyzed their constituents, and published their findings in scientific journals.

King's American Dispensatory is a well-known work representing clinical research into over 600 herbs. This two-volume set was originally written by Dr. John King, Professor of Obstetrics at the Worcester Medical Institute (1869) and President of the Eclectic Medical Association (1878). Upon King's death, John Uri Lloyd revised and updated the volumes. This work is considered to be an herbal literature classic and is still used as a text by many herbalists.

The online text is a scanned version by Henriette Kress (see Chapter 4's entry for Henriette's Herbal Homepage) from the Eighteenth Edition, Third Revision, published in 1898, edited by Harvey Wickes Felter and John Uri Lloyd.

Kumamoto University Medicinal Plant Garden
<http://www.pharm.kumamoto-u.ac.jp/yakusoen/plants-e.html>

This medicinal plant garden is located at Kumamoto University, in Japan, and is affiliated with the school's Department of Pharmaceutical Sciences. Though the garden has over 2,000 plant species, currently only a few images are available online. However, some unusual medicinal plants can be found here, such as the Ayurvedic Indian snakeroot *(Rauwolfia serpentina)* and Chinese pueraria root *(Pueraria lobata)*.

Library of Congress
<http://www.loc.gov/> (home page)

Library of Congress Classification Outline
<http://lcweb.loc.gov/catdir/cpso/lcco/lcco.html>

The LC classification system developed for cataloging LC library materials is still widely used today by the American public and by university libraries. LC subject headings are an example of a "controlled language" or "a controlled vocabulary," and their correct use is usually the most efficient method for retrieving the most relevant records from a database. Browse the RM to RZ subclass section to find relevant headings for searching for herbal medicine materials in the LC and other online catalogs that use this classification system.

Various LC headings are relevant to herbal medicine. Some of the most useful ones are

- MATERIA MEDICA, VEGETABLE;
- MEDICINAL PLANTS;
- BOTANY, MEDICAL;
- HERBAL TEAS; and
- HERBS — THERAPEUTIC USE.

Library of Congress Collections & Services for Researchers, Libraries, and the Public
<http://www.loc.gov/library/>

The U.S. Library of Congress in Washington, DC, was founded in 1800 and is probably the world's largest national library. It primarily served as a congressional library for many years. As the collection grew, through subsequent purchases and as a result of the Copyright Act (one copy of each book copyrighted in the United States is placed in the library), the library gradually became the national library of the United States. It currently houses some 19 million books and 33 million manuscripts; it also holds the largest rare book collection in North America (more than 700,000 volumes). The Library of Congress (LC), however, defers to the National Library of Medicine (NLM) and the National Agricultural Library (NAL) for collecting materials in the fields of clinical medicine and technical agriculture, respectively. Most of LC's collections are open to the public. The library also supports the research community with its outstanding collections of books, manuscripts, music, prints, and maps. It also provides educational programs, with lectures and exhibits.

The library's collections provide a wealth of information for researching herbal medicine. Of particular importance are the following sections and services.

Library of Congress Online Catalog
<http://catalog.loc.gov>

The Library of Congress Online Catalog is a database of approximately 12 million records, representing books, serials, computer files, manuscripts, cartographic materials, music, sound recordings, and visual materials in the library's collections. Searching the catalog with the LC subject heading "Medicinal Plants" retrieved 210 items (see the following discussion of LC subject headings).

The following book titles were retrieved from the library's catalog and show the wide range of herbal materials to be found in the collection:

- *In Praise of Wild Herbs: Remedies & Recipes from Old Provence* by Ludo Chardenon (1984)
- *The Herbal Cures of Duodenal Ulcer and Gall Stones* by Captain Frank Roberts (1969)
- *The South African Herbal* by Eve Palmer (1985)
- *Herb Drugs and Herbalists in Pakistan* by Khan Usmanghani, Gisho Honda, and Wataru Miki (1986)
- *The Fern Herbal: Including the Ferns, the Horsetails, and the Club Mosses* by Elfriede Abbe (1981)

Library of Congress Science Tracer Bullets Online
(SCTB Online)
<http://lcweb2.loc.gov/sctb/>

The Library's Science and Technology Division publishes several subject bibliographies in its "Tracer Bullet" series that serve as useful research guides. These are accessed via the SCTB database (to navigate to this page from the LC home page, select Collections and Services, then Electronic Research Tools, and then Science Tracer Bullets Online; search the collection by typing in relevant terms). Currently, three research guides are of relevance to herbal medicine: *Medicinal Plants*, *Edible Wild Plants,* and *The Ethnobotany of the Americas*.

Library of Congress Vatican Exhibit—Rome Reborn:
The Vatican Library & Renaissance Culture
<http://lcweb.loc.gov/exhibits/vatican/toc.html>

The Vatican Library houses many authoritative works on botany and materia medica. From the opening page of the exhibit, go to the section on Nature for an informative survey of the Greek and Roman founders of herbal materia medica, with sample scanned pages from famous herbals, such as Galen's *De simplicium medicamentorum temperamentis ac facultatibus* and Pliny's *Naturalis Historia.*

The Lloyd Library and Museum
<http://www.libraries.uc.edu/lloyd/>
University of Cincinnati Libraries Catalog
<http://ucolk2.olk.uc.edu/screens/opacmenu.html>

The Lloyd Library in Cincinnati is the largest library of medical plant books in the world and is of major importance to anyone interested in herbalism. It has particularly outstanding collections in the literature of pharmacy and botany, much of it not available elsewhere. John Uri Lloyd (1849-1936) was an important and well-known pharmacist of his time. He was editor of the first *National Formulary* and served twice as president of the American Pharmaceutical Association. With his brothers Nelson Ashley Lloyd and Curtis Gates Lloyd, he founded the company Lloyd Brothers, Pharmacists, Inc., which gained prominence as the manufacturer of a line of botanical products known as "specific medicines." These were designed to meet the needs of the so-called "eclectic" physicians (discussed earlier in this chapter). The Lloyd Brother's collection of books grew into the Lloyd Library and Museum, a 200,000-volume book and periodical collection largely devoted to botany, pharmacy, pharmacognosy, herbal and alternative medicines, natural products, eclectic medicine, and horticulture.

The library provides reference and bibliographic services free of charge. Though it does not loan books or journals from its collections, photocopies of materials are provided through interlibrary loan. The holdings of the library may be searched through the University of Cincinnati Library's electronic card catalog.

A Modern Herbal
<http://www.botanical.com/botanical/mgmh/mgmh.html>

The is the full text of Mrs. Maud Grieve's classic 1931 monograph on the medicinal, culinary, and cosmetic uses and folklore of herbs, with brief information on their habitats and their geographical distribution (for detailed information, see the corresponding entry in Chapter 7).

Southwest School of Botanical Medicine
<http://chili.rt66.com/hrbmoore/HOMEPAGE/HomePage.html>

The Southwest School of Botanical Medicine, located in Bisbee, Arizona, is a private institution providing training programs for the herbalist. Its founder and director, Michael Moore, is a respected U.S. herbalist and the author of several books on medicinal plants. This Web site is one of the most outstanding and interesting sites for information on herbal medicine, providing a wealth of information geared toward the traditional herbal practitioner. Among the highlights are the full text of classic nineteenth- and twentieth-century texts, such as the Lloyd Brothers plant drug pamphlets (1897 to 1915); Fyfe's *Materia Medica; The Eclectic Materia Medica, Pharmacology and Therapeutics* by Harvey Wickes Felter (1922); and *Culbreth's Materia Medica,* which deals with all medicinal plants that were ever official entries in the USP and the NF.

Chapter 3

Identification, Cultivation, and Conservation: Botanical and Agricultural Resources

A man may esteem himself happy when that which is his food is also his medicine.

Henry David Thoreau

WHAT IS AN HERB?

What exactly is an herb? The term *herb* can mean different things to different people, with definitions varying according to the user's area of interest and personal bias. The word *herb* itself is derived from the Sanskrit *bharb,* meaning to eat. This in turn became the Latin word *herba,* meaning grass or fodder, and, in Middle English, *herbe*. The strict botanical definition is "a seed-producing annual, biennial, or perennial that does not develop persistent woody tissue but dies down at the end of a growing season" (*Merriam-Webster's Collegiate Dictionary,* Tenth Edition). This definition, however, would exclude important "herbs" such as hawthorn (which is a shrub) and willow (a tree). A more satisfactory definition now generally in use is the second one given in *Merriam-Webster's Collegiate Dictionary:* "a plant or plant part valued for its medicinal, savory, or aromatic qualities."[1]

PLANT NAMES AND PLANT TAXONOMY

Information resources usually list herbs by their common or botanical names. A common name can be in any language, is usually based on what the flower or plant looks like, and will vary according to where you happen to live: for example, St. John's wort is also known as "Le millepertuis" in France and "Johanniskraut" in Germany. To circumvent the problem of multiple common names, each plant is given a botanical name (also called the scientific or Latin name) based on an internationally accepted system for naming each unique and distinct plant, whether natural or cultivated. Though plants may have several common names, they usually have only one botanical name. The botanical name is in Latin, is italicized or underlined, and is "binomial," that is, has two parts: the first name indicates the genus (plural, genera), a group of closely related plants; the second name designates the species, the specific name for the particular type of plant. Botanical names have the distinct advantage of referring to only one plant. For example, the botanical name for the most commonly used species of echinacea is *Echinacea purpurea,* while two of its common English names are black sampson and coneflower. When creating a botanical name, botanists also usually add the name of the person who first named the plant, so, for example, the botanical name of garlic is *Allium sativum* L., the "L" standing for Linnaeus. Botanical names become valid only when they are published according to specific rules, with illustrations and technical descriptions of the plant.

Some of the resources in this chapter organize entries by botanical family or some other taxonomic category. Taxonomy, or systematics, is the science of classifying plants and animals, grouping species into a hierarchy of categories based on their genealogy, or family tree, and not necessarily their physical similarities. Related species are grouped into genera, genera into families, families into orders, and so on. For example, the main taxonomic categories for echinacea are, from broad to narrow, kingdom Plantae; division Spermatophyta; class Angiospermae; subclass Dicotyledonae; family Asteraceae; genus *Echinacea.*

Some resources will also list the pharmacopoeial name, perhaps only widely used in European countries. This is a convenient system for identifying a substance by referring to its scientific name plus the plant part or type of preparation being used. Some herbs will thus be associated with more than one pharmacopoeial name: preparations made from the root of *Echinacea* are referred to as Echinacea radix, while those using the aerial parts of the plant are known as Echinacea herba. (Table 3.1 lists some of the different names the user will encounter in the literature for the herb valerian.)

THE IMPORTANCE OF CORRECT PLANT IDENTIFICATION

The authors of a rather famous 1969 article published in the distinguished *Journal of the American Medical Association* (JAMA) suggested that catnip *(Nepeta cataria),* when smoked, produced a psychedelic high not unlike cannabis.[2] It was later discovered that the researchers had, in fact, mistakenly used the cannabis plant instead.[3] Though we can assume the authors of the 1969 paper were appropriately embarrassed, their report does nicely illustrate the importance of proper botanical identification. Some plants are almost indistinguishable from others that have totally different properties, and this can have fateful consequences. One example of serious misidentification occurred in 1997 when someone taking an herbal product containing plantain (*Plantago major* L.) developed life-threatening complications. It was subsequently discovered that the supply of what was supposed to be plantain was contaminated with foxglove *(Digitalis lanata),* a plant containing powerful heart stimulants: inexperienced herb collectors in Europe had apparently confused the two plants.[4]

Herbs are often misidentified, or the wrong part of the plant may be harvested. For example, echinacea refers to a genus of nine types of plant native to North America, including *Echinacea angustifolia, E. purpurea,* and *E. pallida.* Commercial supplies of echinacea root are

TABLE 3.1. Names Used for Valerian Root/Rhizome[1]

Latin/scientific/botanical name	*Valeriana officinalis* L.
Common names	Garden valerian, Mexican valerian, garden heliotrope, Indian valerian, All-Heal
Pharmacopoeial name[2]	Valerianae radix

[1]A rhizome looks like a root but botanically is really part of the stem.
[2]Pharmacopoeias provide official standards of identity, purity, preparation, etc.

commonly adulterated by another plant, *Parthenium integrifolium* (wild quinine), mistakenly identified as *E. purpurea*.[5] Note also that closely related species, and even parts of the same plant, often have different pharmacological activities. For example, two parts of the echinacea plant are used as remedies: the herb, that is, the above-ground portion, including flowers, and the root. Germany's Commission E monographs approve the use of *E. pallida* root and *E. purpurea* herb but find no evidence to support the use of *E. purpurea* root and *E. angustifolia* root.[6]

The following Web sites are online resources providing basic botanical information, plant images, plant taxonomic data, plant identification, conservation issues, and information about the cultivation of herbs.

WEB SITES PROVIDING BOTANICAL, CONSERVATION, AND CULTIVATION DATA

Agribusiness in Sustainable Natural African Plant Products (A-SNAPP)
<http://www.herbs.org/africa>

Many important medicinal plants are native to South Africa, such as buchu *(Barasoma betulina),* traditionally used in the treatment of stomach ailments and rheumatism. A-SNAPP is a collaborative project of the Herb Research Foundation (HRF) and the Agricultural Research Council (ARC) of South Africa. It is dedicated to the development of a "socially conscious and environmentally sustainable" African natural products business.

Brooklyn Botanic Garden
<http://www.bbg.org/gardening/>

The gardening section of the Brooklyn Botanical Garden (BBG) Web site offers a wealth of basic botanical information designed for the nonspecialist. At the home page scroll down to the Garden Botany section for articles on such topics as plant names, plant classification systems, plant structure and function, plant conservation, and other issues.

The CalFlora Database
<http://elib.cs.berkeley.edu/calflora/>

CalFlora is a collaborative project of the USDA Forest Service, the UC Berkeley Digital Library Project, the U.S. Geological Survey, the UC Davis Information Center for the Environment, and the Santa Barbara Botanic Garden. This comprehensive database of vascular plant species known to occur in California is one of the most sophisticated electronic resources for plant information on the Internet. It provides resources for research on plant distribution and diversity patterns; information about individual species, such as habitat, names, and rarity status; as well as links to other online resources, including maps, pictures, specimen records, and other botanical database systems. It contains geographic and ecological distribution information for 8,363 vascular plant groups in California, as well as additional habitat information for rare species of the Sierra Nevada. Though obviously restricted to plants in California, it provides much useful information on several native species that have used for medicinal purposes, such as Mormon tea *(Ephedra nevadensis),* used by American Indians and early settlers to treat kidney and stomach ailments, and blue elderberry *(Sambucus mexicana),* used for a variety of ailments, including the treatment of wounds.[7] Figure 3.1, shows part of the CalFlora record for St. John's wort *(Hypericum perforatum).*

CalPhotos: Plants (California Plants and Habitats)
<http://elib.cs.berkeley.edu/photos/flora/>

CalPhotos, part of the Berkeley Digital Library Project, is one of the largest collections of plant photographs on the Internet, with over 20,000 photos of California plants from the Brousseau Collection, the California Academy of Sciences, California state agencies, and others. Many of these plants have medicinal uses. This site may be browsed by common or scientific name. It has links to the related CalFlora database (see previous entry).

Convention on International Trade in Endangered Species
of Wild Fauna and Flora (CITES)
<http://www.cites.org/CITES/>

Many species of plants and animals are declining in number because of loss of habitat and increased exploitation as human populations grow. The worldwide surge of interest in herbal medicine has

FIGURE 3.1. Part of the CalFlora Record for St. John's Wort *(Hypericum perforatum)*

CalFlora Taxon Report

Hypericum perforatum L. (Hypericaceae)

Common names: Klamath weed [Hrusa], common St. Johnswort [PLANTS]
Plant communities: weed, species characteristic of disturbed places [Lum/Walker]
Habitat: described by Walker and/or CNPS as occurring in disturbed and disturbed habitats [Walker and/or CNPS Inventory 1994]
Elevation: between 0 and 4500 feet [Lum/Walker]

☐ Brother Alfred Brousseau (1 of 1)

Hypericum perforatum, a dicot in the family Hypericaceae, is a perennial herb that is **not native** to California [Hrusa]; it was introduced from elsewhere and naturalized in the wild [Lum/Walker].

The California Exotic Pest Plant Council (CalEPPC) lists Hypericum perforatum on List B: Wildland pest plant of lesser invasiveness; causes lesser degree of habitat disruption than plants on list A. [CalEPPC]
Hypericum perforatum is also classified by the California Department of Food and Agriculture as Noxious Weed List C: Control required in nurseries, not required elsewhere. [CDFA Weeds 2000]

Source: The CalFlora Database [Web application]. 2000. Taxon report for *Hypericum perforatum.* <http://www.calflora.org/cgi/calflora_query?special=calflora &where-calrecnum=4312&one=T> [accessed January 11, 2001].

been accompanied by an increase in demand for plant material, and many of the most widely used herbs are being overharvested from their natural habitats (see also the entry for United Plant Savers). In 1973, a treaty was drawn up to protect both fauna (animals) and flora (plants) from overexploitation and to prevent international trade from threatening species with extinction. This treaty, the Convention on International Trade in Endangered Species of Wild Fauna and Flora (CITES), was signed in Washington, DC, on March 3, 1973, and came into force on July 1, 1975. The 152 member countries act by banning commercial international trade in animals and plants included on an agreed-upon list of endangered species, and by regulating and monitoring trade in others that might become endangered.

At a recent CITES meeting, the convention accepted a U.S. proposal to include the threatened medicinal herb goldenseal *(Hydrastis canadensis)* on its Appendix II list: commercial trade in goldenseal may continue but is now subject to certain regulations and requires a permit.

CITES maintains a large database, holding some 3.3 million records on trade in wildlife species and their derivative products. The information spans from 1975 to the present and is constantly being updated. In addition to the trade records, the database holds some 30,000 scientific names and synonyms. To access the database, go to the English-language section, then select Flora from the CITES database menu; search by name or country.

The Families of Flowering Plants
<http://biodiversity.uno.edu/delta/angio/>

The correct identification of plants is important, especially because some herbs are almost indistinguishable from plants having totally different properties. This database is an authoritative resource for information on the description and identification of the flowering-plant families, also known as angiosperms, a category that includes fruits, vegetables, herbs, and spices. The descriptions are detailed, including information about habit and leaf form; leaf and stem anatomy; reproductive type and pollination; flower, fruit, and seed morphology; seedling germination; physiology and biochemistry; geography and cytology; and taxonomic placement. Most entries include illustrations and are supplemented with literary quotations. The Families of Flowering Plants, created and maintained by L. Watson and M. J. Dallwitz, follows the DELTA system (a system for processing taxonomic descriptions: DEscription Language for TAxonomy) in the Taxonomy Laboratory of the Research School of Biological Sciences at the Australian National University.[8] As the title of this resource suggests, information in this database is organized by plant family. To find information on plants in the genus *Echinacea,* for example, you would look up the entry for Compositae (also known as Asteraceae), the daisy family.

Flora Celtica Database
<http://www.rbge.org.uk/research/celtica/dbase/searchform.html>

Flora Celtica is a project seeking to gather information on the role of plants in the lives of the people of Scotland and Celtic Europe (see the corresponding entry in Chapter 7).

Flora of China
  <http://flora.huh.harvard.edu/china/>

Mainland China has about 30,000 plant species, or one-eighth of the world's total number, including about 8,000 species of plants that are of medicinal or economic importance. Many of these plants are vital components of the Chinese materia medica, part of a system of traditional medicine stretching back thousands of years (see Chapter 7). Flora of China is a collaborative international project designed to publish the first modern English-language account of these plants. Selected information from these volumes can be viewed at this Web site. Images from the volumes can be viewed through the Missouri Botanical Garden's VAST (VAScular Tropicos) nomenclature database (see entry in this chapter for Missouri Botanical Garden).

Flora of North America (home page)
  <http://hua.huh.harvard.edu/FNA/>

Flora of North America (online database)
  <http://hua.huh.harvard.edu/cgi-bin/Flora/flora.pl?
  FLORA_ID=12395>

This cooperative effort between Canadian and American botanists aims to catalog the estimated 21,000 native plants on the North American continent, including information on their medicinal uses. Though this information is being published in several print volumes, various data can be viewed by selecting The Flora at the home page, then clicking on Search to access the database search screen. To view a sample record, for example, type "dog nettle" in the text box; this will bring up the entry for *Urtica urens,* a plant used by some American Indian tribes to treat rheumatism.[9]

GardenGuides
  <http://www.gardenguides.com/>

This is another Web site intended for the lay gardener, but with valuable information on specialized herbal topics, such as wildcrafting (the gathering of plant material from its native environment), herb cultivation, and herb harvesting and drying. The site provides useful guide sheets for most of the popular herbs, focusing on their cultivation. Of special interest is the section Herbs and Pregnancy.

The Garlic Information Centre
<http://www.mistral.co.uk/garlic/>

The Garlic Information Center, based in East Sussex, England, was created by the chemist and author Peter Josling and has useful information on garlic and its medicinal use. This site posts research abstracts concerning garlic's pharmacological properties and clinical activity.

Herbs and Herb Gardening: An Annotated Bibliography
and Resource Guide
<http://www.nal.usda.gov/afsic/AFSIC_pubs/srb9606.htm>

This is an extensive annotated bibliography on the history and cultivation of herbs compiled by Suzanne DeMuth at the National Agricultural Library (NAL)'s Alternative Farming Systems Information Center (AFSIC) (see the entry for the National Agricultural Library in this chapter). It includes resources on herb gardens and home gardening with herbs (see also Chapter 2).

Interactive European Network for Industrial Crops
and their Applications (IENICA)
<http://www.csl.gov.uk/ienica/>

The IENICA project is funded by the European Commission, the policymaking arm of the European Union (EU). It seeks to identify and create scientific, industrial, and market opportunities for specific industrial crops or applications. IENICA involves fourteen European states and draws expertise and experience from academia, industry, and agriculture. IENICA database provides comprehensive background information on medicinal herbs grown in individual European countries, including data on cultivation, uses, and the effect of growing conditions on a plant's active compounds. Contact information is also provided for European universities and organizations currently involved in research.

Köhler's *Medizinal Pflanzen* (Köhler's Medicinal Plants)
<http://www.mobot.org/MOBOT/research/library/kohler/
welcome.html>

The Hermann A. Köhler's *Medizinal Pflanzen,* collection of medicinal plant illustrations, was published in 1887 but is considered by many to be one of the finest and most useful series of illustrations of medicinal plants. These images from the collection are made available by the Missouri Botanical Garden as part of its online rare books project. (See entry for Missouri Botanical Garden in this chapter.) Plants are listed by scientific name, and synonyms or common names are not included. To see a sample illustration, select Text Listing of Scientific Names from the home page, then click on any entry.

Laboratories for Natural Products, Medicinal, and Aromatic Plants
   <http://www-unix.oit.umass.edu/~herbdig/>

Laboratories for Natural Products, Medicinal and Aromatic Plants, part of the Department of Plant and Soil Sciences at the University of Massachusetts, Amherst, is "an organization of faculty, staff, and students on the University of Massachusetts at Amherst campus" that provides "research, consulting, and teaching services in a number of areas related to natural products, medicinal and aromatic plants." Current areas of research include medicinal herbs of China and Russia. Publications include the newsletter *Herb, Spice and Medicinal Plant Digest,* a useful resource containing concise reports on the production, chemistry, and marketing of herbs, spices, and medicinal plants. A paid subscription is required for the newsletter, but the table of contents and selected pages can be viewed online for free.

The LuEsther T. Mertz Library (New York Botanical Garden)
   <http://www.nybg.org/bsci/libr/>

The LuEsther T. Mertz Library of the New York Botanical Garden is one of the world's largest and most active botanical and horticultural libraries. It collects published literature in such areas as economic botany, plant ecology, horticulture and gardening, landscape design, garden history, and botanical and horticultural bibliography. Its online catalog is CATALPA, the CATAlog for Library Public Access (also a small genus of American and Asiatic trees of the trumpet-creeper family Bignoniaceae). The library provides reference and information services to commercial, academic, and scientific users, as well as the general public.

Missouri Botanical Garden
  <http://www.mobot.org/>

Missouri Botanical Garden's VAST (VAScular Tropicos) Nomenclature Database
  <http://www.mobot.org/W3T/Search/vast.html>

The Missouri Botanical Garden, in St. Louis, Missouri, is an important center for botanical research and educational activities. This Web site has become one of the Internet's largest and most frequently used repositories of botanical information in the world. Of note is the garden's Applied Research Department, members of which are collaborating with pharmaceutical researchers in the search for new anti-HIV and anticancer medicines (to navigate to this section, select Research on the home page). It also has one of the world's finest botanical libraries.

Click on W^{3}Tropicos to access the VAST (VAScular Tropicos) nomenclature database, a gateway to a large collection of resources, including cytological studies (e.g., number of chromosomes), specimen lists, and distribution maps, along with bibliographic references and plant images.

National Agricultural Library (NAL)
  <http://www.nalusda.gov/>

The National Agricultural Library (NAL) is one of four national libraries in the United States: the others are the Library of Congress, the National Library of Medicine, and the National Library of Education. The primary source for agricultural information in the United States, the library seeks to "increase the availability and utilization of agricultural information for researchers, educators, policymakers, consumers of agricultural products, and the public." It also serves as the U.S. center for coordinating international access to resources and information.

NAL produces AGRICOLA (AGRICultural OnLine Access), a bibliographic database with citations on all aspects of agriculture and allied disciplines, including food and nutrition (for information on searching this database, see the corresponding entry in Chapter 13).

The library provides several resources of relevance to herbal medicine. Of particular interest here are the Quick Bibliographies. At the home page, select Publications and Databases, then Bibliographies and Other Reference Publications; go next to the Alternative Farming Systems (AFSIC) section to select particular bibliographies, such as *Herbs and Herb Gardening, Growing for the Medicinal Herb Market,* and *Resource Guide to Growing and Using Herbs.*

The National Center for the Preservation of Medicinal Herbs <http://www.ncpmh.org/>

One unfortunate consequence of the recent resurgence of interest in herbal medicine is the overharvesting of plants from their native habitats. The National Center for the Preservation of Medicinal Herbs is a not-for-profit research facility and preserve located in Meigs County, Ohio. It cultivates and studies medicinal herbs at risk of extinction due to excessive harvesting, such as bloodroot *(Sanguinaria canadensis)* and goldenseal (*Hydrastis canadensis* L.). This site provides valuable information on cultivation and preservation issues, including current research projects designed to tackle these problems (also see the related entry in this chapter for United Plant Savers).

National Technical Information Service (NTIS) <http://www.ntis.gov/>

The National Technical Information Service (NTIS) provides a centralized service for the distribution of scientific, engineering, technical, and related business information produced for or by the U.S. government. The NTIS catalog provides access to over 400,000 publications plus CD-ROMs, datafiles, and audiovisuals, searchable by subject or keyword. Full text is not provided, but summaries are included and documents may be ordered from the NTIS for a fee.

Searching on the keywords "medicinal herbs" retrieves a wide range of useful reports, including information on herbal medicine use and production in foreign countries. Representative titles include *Guidelines for the Chemical and Biological Assessment of Herbals and Herbal Preparations; Factory-Produced Herbal Medicine—The Indian Experience; Production of Pharmaceutical Materials from Medicinal and Aromatic Plants; Turkey;* and *Guidelines for Setting Up Data Bases on Medicinal Plants.*

NatureServe
<http://www.natureserve.org/>

The NatureServe Web site is provided by the Association for Biodiversity Information (ABI), in collaboration with the Natural Heritage Network. The database is a source for authoritative conservation information on more than 50,000 plants, animals, and ecological communities of the United States and Canada, providing conservation status, taxonomic and distribution information, and references. It includes a listing of all U.S. states where a particular plant has been identified and the species conservation status.

NewCROP: New Crop Resource Online Program
<http://newcrop.hort.purdue.edu/newcrop/>

NewCROP Guide to Medicinal and Aromatic Plants
<http://newcrop.hort.purdue.edu/newcrop/med-aro/>

This Web site for the Center for New Crops and Plant Products at Purdue University, West Lafayette, Indiana, is associated with the New Crop Diversification project and the Jefferson Institute. The center's mission is to identify new natural sources of industrial or medically useful compounds and to identify novel compounds from plants. Research projects on the cultivation of aromatic and American medicinal plants, such as mint, ginseng, goldenseal, cohosh, and echinacea are also currently underway. Resources such as CropINDEX, CropSEARCH, CropMAP, and CropREFERENCE provide access to a wide range of information and documents on various crop plants, including some economically important herb crops. Many of the documents are available as online full text.

The NewCROP Guide to Medicinal and Aromatic Plants section of the site is specifically designed for access to information resources on aromatic and medicinal herbs (select Aromatic-MedicinalPLANTS on the home page). General monographs provide background information about herb plants, a listing of herb varieties available from commercial sources, a guide to public and commercial sources of these plants, and a searchable database of companies in the botanical products industry.

Plants Database
  <http://plants.usda.gov/plants/>

This project of the U.S. Department of Agriculture (USDA) Natural Resources Conservation Service is intended to be a single source of standardized information on the vascular plants, mosses, liverworts, hornworts, and lichens of the United States and its territories. The database includes names, identification information, distributional data, crop information, plant growth data, references, and other plant information. Entries can be retrieved by both common and botanical names.

Southwest School of Botanical Medicine
  <http://chili.rt66.com/hrbmoore/HOMEPAGE/HomePage.html>

This Web site has one of the most extensive collections of medicinal plant images available on the Internet, around 1,700 photographs, line drawings, and woodprints from a variety of sources, organized by genus. It also provides distribution maps for medicinal plants found in the western United States (see the corresponding entry in Chapter 2 for information on other types of resources available on this Web site).

United Plant Savers (UpS)
  <http://www.plantsavers.org/>

The explosive growth of the international herbal market is creating such a demand for herbal plant material that some medicinal plants are now in short supply due to overharvesting in the wild. United Plant Savers (UpS) is a grassroots nonprofit organization group based in East Barre, Vermont, whose mission is to promote the conservation of North American native medicinal plants. It also maintains its own sanctuary for research and cultivation on 400 acres in southeast Ohio. According to UpS, twenty Native American herbs, such as black cohosh *(Cimicifuga racemosa)* and American ginseng *(Panax quinquefolius),* are "at risk," that is, are approaching endangered species status, while another nineteen are on a "to watch" list.

University of Washington Medicinal Herb Garden
  <http://www.nnlm.nlm.nih.gov/pnr/uwmhg/>

The University of Washington's Medicinal Herb Garden was first planted in 1911 by the university's dean of pharmacy. The 2.5-acre garden is the largest of its kind in the United States and was designed to serve as a source of information on the uses of herbaceous plants in traditional medicine and in the home and garden. The garden contains around 300 varieties of plants. One section focuses on medicinal plants of the Pacific Northwest, such as cascara trees *(Rhamnus purshiana),* while other sections feature plants from around the world.

This Web site was developed by Michael Boer and staff of the National Network of Libraries of Medicine, Pacific Northwest Region (NN/LM PNR), one of eight regional medical libraries sponsored by the National Library of Medicine (NLM). It is designed to be a hypertext tour of the garden, though it is only a "partial catalog" and some common herbs will not be found here. Records can be accessed by both common and botanical names. Each entry includes photographs of the plant, plus links to corresponding entries in other databases, such as the Plants Database (see entry in this chapter) and EthnobotDB, an ethnobotany database (see the corresponding entry in Chapter 7).

Of particular note is the link to the National Library of Medicine's (NLM) PubMed database, containing MEDLINE citations (see the entry for PubMed in Chapter 13). Simply click on the PubMed link to retrieve recent journal citations for that particular herb.

United States Department of Agriculture (USDA)
  <http://www.usda.gov/>

USDA Agricultural Research Service (ARS)
  <http://www.ars.usda.gov/>

USDA-ARS Natural Products Utilization Research Unit
  <http://www.olemiss.edu/depts/usda/>

The USDA was established by President Abraham Lincoln in 1862. Its principal mission is to support American farmers and American agriculture. Some of its various offices and programs directly concern the cultivation and use of herbs. For herbal medicine, the most pertinent sections of the USDA are the National Agricultural Library (NAL, discussed earlier in this chapter) and the Natural Products Utilization Research Unit, working to promote natural products for agriculture and to support the development of medicinal plants as alternative crops. The well-known author and botanist James Duke was employed by the USDA for almost thirty years, and the ARS still hosts his Phytochemical database (see Chapter 13).

Other miscellaneous reports and documents are to be found buried in this site, for example, statistical reports on the cultivation of herbs in Hawaii, published by the National Agricultural Statistics Service (NASS). To search the USDA site, type keywords into the search box at the top of the home page.

Vascular Plant Image Gallery
<http://www.csdl.tamu.edu/FLORA/gallery.htm>

This searchable gallery of 8,800 photo images of native North American plants is provided by Texas A&M University Bioinformatics Working Group.

Chapter 4

The Traditional Herbalist Movement and the U.S. Herbal Industry

Tradition is a guide and not a jailer.

W. Somerset Maugham

THE HERBALIST APPROACH TO HEALTH

The term *herbalism* has changed over the years, once being almost synonymous with botany, though it is now generally used to refer to the use of plant remedies to treat medical conditions. It is often associated with medical systems having a tradition of using herbs for healing, whether in North America, Europe, or China. "Western herbalism," as practiced in the United States and Europe, and distinct from allopathic medicine, is characterized by an emphasis on building and maintaining health rather than fighting disease. More significant, traditional herbalists practicing outside mainstream medicine often view healing very differently from allopathic physicians. The latter tend to regard herbal preparations as being essentially no different from conventional pharmaceutical drugs. Many herbalists take a more holistic approach, believing that "herbal medicine is a healing technique that is inherently in tune with nature."[1] This emphasizes the self-healing capacity of the body and its supposed activation by herbs. Many herbalists also insist on the superiority of "natural" drugs over their synthetic equivalents, a view that is particularly inimical to mainstream scientific medicine.

Much of the traditional herbal literature has a strong mystical component, and this can be a little unsettling for those health practitioners approaching this literature for the first time. Herbal writings often touch upon New Age theology and astrology, and useful information sometimes is buried among discussions of "pagan" religious practices and the association of different herbs with the various planets in the solar system. However, it should be remembered that before the current entry of herbs into mainstream medicine, a considerable body of herbal knowledge was kept alive by individuals and movements operating on the fringes of orthodox medicine. Some of the most widely used information resources on the Internet are provided by individuals who are not graduates of mainstream medical or pharmacy schools.

HERBAL PRODUCTS AND MANUFACTURERS

Along with the mainstreaming of herbal medicine in the United States has come a burgeoning of the market for herbal preparations. The actual size of the herbal market in the United States is difficult to determine, but it is now believed that annual sales exceed $3 billion.[2] Herbal preparations are moving beyond health food stores into pharmacies and grocery stores. Much of the sales growth can be traced to the passage of the Dietary Supplement Health and Education Act in 1994, which critics argue allowed herbal companies to promote their products with little oversight, as long as they do not claim that the products treat a specific disease or condition. The more responsible members of the herbal industry, though, including many manufacturers of herbal preparations, realize that if herbal medicines are to be integrated successfully into the health care system, companies will have to deal with issues such as safety, toxicology, and interactions with conventional drugs. Supported by the herbal trade associations, such organizations as the American Herbal Products Association (AHPA) are beginning to tackle problems such as the development of quality control standards for the manufacture of herbal supplements. AHPA is currently developing monographs for an American herbal pharmacopoeia that is intended to be a primary reference tool for health care providers, manufacturers, and regulators. It will include many of the Ayurvedic, Chinese, and Western herbs most frequently used in the United States.

This chapter includes those Internet sites which are maintained and most frequently used by practicing herbalists in the United States. Also included are sites administered by or affiliated with groups or organizations serving the U.S. herbal industry and dietary supplement manufacturers.

WEB SITES OF PRACTICING HERBALISTS AND U.S. HERBAL ORGANIZATIONS

Algy's Herb Page
<http://www.algy.com/herb/>

Algy's Herb Page was one of the Web's first sites dedicated to herbs and their many uses. A personal Web site run by the anonymous "Algy," it is essentially a gateway to a series of electronic bulletin boards where herb enthusiasts can exchange ideas. This well-known Web site provides unmonitored, Web-based, open discussion groups dedicated to the exchange of information on the history, folklore, wildcrafting, growing, harvesting, and uses of culinary and medicinal herbs. Select Apothecary at the home page to enter the herbal medicine section. Posts remain on the Web site for about one week and are then deleted as new threads arise.

The American Herbal Pharmacopoeia (AHP)
<http://www.herbal-ahp.org/>

A pharmacopoeia lists and provides information on medicinal drugs, including data on their preparation and testing methods for purity. The AHP is intended to be just such a pharmacopoeia for the herbal industry, consisting of quality authoritative monographs with accurate, critically reviewed information. It will provide guidance in the appropriate use and preparation of herbal products, as well as encouraging the development of quality control standards for herbal supplements and botanical medicines. Writers and contributors include prominent medical herbalists from America, France, India, China, and Australia.

Eventually the AHP will include about 300 monographs. AHP contributors are currently working on thirty monographs that may be purchased separately as soon as each one has been completed. So far monographs have appeared for St. John's wort *(Hypericum perforatum);* valerian root *(Valeriana officinalis);* hawthorn flower, leaf, and berry *(Crataegus* spp.); astragalus root *(Astragalus membranaceus);* schisandra berry *(Schisandra chinensis);* and willow bark *(Salix* spp.). Extracts from the monograph for St. John's wort can be viewed by selecting "AHP Monographs" from the title page.

The American Herbal Products Association (AHPA)
<http://www.ahpa.org/>

The American Herbal Products Association (AHPA) was founded in 1983 by a group of herbal product companies to promote the responsible commerce of products containing herbs. It is considered to be the national trade association and voice of the herbal products industry, composed of companies doing business as growers, importers, manufacturers, and marketers of herbs and herbal products.

AHPA publishes two important books. The *Botanical Safety Handbook* provides safety data for more than 600 commonly sold herbs, including international regulatory status, standard dosage, and common toxicity problems. *Herbs of Commerce,* a reference text designed to address confusion about the common names of botanical ingredients, attempts to standardize botanical and common names for herbs. The original edition was adopted as the FDA's standard source for naming botanical ingredients on dietary supplement labels.

Go to the News section for up-to-date items concerning legislation affecting the manufacture and marketing of herbs in the United States.

Consumer Healthcare Products Association (CHPA)
<http://www.chpa-info.org/>

The CHPA is a national trade association representing manufacturers of nonprescription, over-the-counter (OTC) medicines and dietary supplements. Go to the Issues section for information and statistics on dietary supplements.

Council for Responsible Nutrition (CRN)
<http://www.crnusa.org/>

The Council for Responsible Nutrition (CRN) is a trade association of the nutritional supplements, ingredients, and other nutritional products industry. It provides its member companies with legislative guidance, regulatory interpretation, and scientific information on issues relating to dietary supplements. The CRN played a key role in the passage of the Dietary Supplement Health and Education Act (DSHEA) of 1994 (see Chapter 8 for a discussion of DSHEA).

Dietary Supplements: An Advertising Guide for Industry
<http://www.ftc.gov/bcp/conline/pubs/buspubs/dietsupp.htm>

The Federal Trade Commission (FTC) has responsibility for monitoring the advertising of dietary supplements in print and broadcast advertisements, infomercials, catalogs, and similar direct marketing materials. Such advertising must be truthful, substantiated, and not misleading. The FTC has taken action against supplement manufacturers, advertising agencies, distributors, retailers, catalog companies, and others involved in the deceptive promotion of herbal products. (See Chapter 11 for more information on the role of the FTC.)

European Herbal Practitioners Association (EHPA)
<http://www.users.globalnet.co.uk/~ehpa>

The European Herbal Practitioners Association was founded in 1993 to represent professional herbal practitioners in the European Union (EU). One of its main objectives is to protect both herbalists and consumers by helping to develop professional standards for herbal education and practice, and by encouraging appropriate European legislation. This site primarily provides information on issues relating to political and legislative developments throughout the EU in regard to herbs and the practice and licensing of herbal practitioners. Current issues include the legal status of herbal medicines in the EU and the right of herbal practitioners to practice as autonomous health providers.

Global Health Calendar: Meetings and Other Events
for Herbal Medicine
<http://www.healthy.net/calendar/index.asp>

Provided by HealthWorld Online (see the corresponding entry in Chapter 12 for more detailed information about HealthWorld Online), this comprehensive resource provides information on upcoming meetings and courses around the world in all areas of complementary and alternative medicine (CAM). This site features information on upcoming events such as trade shows, professional education events, and industry seminars. For meetings on herbal topics, select appropriate categories from the Event Topic menu. Herb-related events can be found under several headings, including Acupuncture/Oriental Medicine, Neutraceuticals, Nutrition, and Herbs and Herbalism.

Healthwell.com's Healthwell Exchange
<http://www.healthwellexchange.com/>

The Natural Foods Merchandiser
<http://www.healthwellexchange.com/nfm-online>

Nutrition Science News
<http://www.healthwellexchange.com/nutritionsciencenews/>

This site is designed to be an Internet information network for manufacturers of natural products, such as herbal supplements. It contains up-to-date news, feature articles, and current market research information. Searching by keyword or category will provide access to a comprehensive database of targeted feature articles for individual sectors of the industry.

From the home page click on the Retailing button, then go to the Online Magazines section to access *The Natural Foods Merchandiser* and *Nutrition Science News* (NSN). The former provides extensive information on the herbal industry in the United States. The latter is the digital counterpart of *Nutrition Science News,* a monthly trade magazine written by health professionals. NSN is a useful source of information on vitamins, supplements, herbs, homeopathy, and nutraceuticals.

Henriette's Herbal Homepage
<http://www.ibiblio.org/herbmed/>

This popular and comprehensive personal site began life as a list of frequently asked questions (FAQs) for the newsgroup <alt.folklore. herbs>; it was also one of the earliest Internet sites where herbalists could exchange information (for details on online communication options, see the corresponding entry in Chapter 16). This site is devoted to both medicinal and culinary herbs and is maintained by Henriette Kress, a well-known herbalist located in Helsinki, Finland. In addition to bulletin boards, it includes plant images, information on herbal software, files with herb common names in several languages, and the full text of extracts from such classic herbal manuals as *Specific Medication and Specific Medicines* (1870) by John Scudder, *King's American Dispensatory* by Harvey Wickes Felter and John Uri Lloyd (1898, Eighteenth Edition, Third Revision), and volumes of *The Eclectic Medical Journal* (see Figure 4.1).

The Herbal Bookworm
<http://www.pond.net/~herbmed/>

This collection of book reviews was created by Jonathan Treasure, Member of the National Institute of Medical Herbalists (MNIMH), a practicing medical herbalist. He is a member of the American Herbalists Guild (AHG) and contributor to the *American Herbal Pharmacopoeia* (AHP). The site contains informal but informative reviews of books on medicinal herbs and herbalism, primarily intended for physicians, pharmacists, and allied health care professionals. Note, however, that Mr. Treasure is the author of all of these reviews.

The Herbal Encyclopedia
<http://www.wic.net/waltzark/herbenc.htm>

This is another well-established and well-known personal Internet site for herbal information, but with a strong theological or mystical

FIGURE 4.1. Henriette's Herbal Home Page: One of the Most Popular Herbal Internet Sites

Henriette's herbal homepage - herbfaqs - neat stuff - classic texts - archives - links I like - pictures - plant names

Henriette's Herbal Homepage

Last updated 04Jan01 - email comments to hetta@saunalahti.fi.

This site is now
http://ibiblio.org/herbmed
It's just another namechange. This one should be stable for at least five years. Please update your links and bookmarks.

Check out my guestbook.
(Older entries here: guestbook-1.html)

slant: the entry for echinacea, for example, provides the information that it "is used as an offering to the spirits or gods and goddesses to strengthen a spell or ritual." It is intended as a quick reference guide to the most commonly used American herbs and is maintained by Lisa Waltz, a certified naturopathic doctor (see the entry for the American Society of Naturopathic Physicians in Chapter 5) and "ordained minister." It provides useful brief information for the most widely used herbs, including data on their cultivation and religious symbolism.

Herbal Hall
<http://www.herb.com/>

This site has a comprehensive collection of articles and news items for practicing American herbalists, including an Herb Walk with photographs and other images of herbs. It also hosts newsgroups for professional herbalists (see the corresponding entry in Chapter 16).

HerbWeb
<http://www.herbweb.com/>

This site was created by Tim Johnson and is based on his book *CRC Ethnobotany Desk Reference.*[3] It provides information about 21,000 plant species and the ailments they have traditionally been used to treat. Data include chemical content, nutrition information, and worldwide distribution. The site also includes a complete botanical inventory of twenty-eight national parks and an index of indigenous groups who have used the herbs. Information can be accessed by ailment, action (e.g., analgesic), common and botanical names, family name, use, indigenous use, and geographical range.

Howie Brounstein's Home Page
<http://www.teleport.com/~howieb/howie.html>

This is another well-known personal herbal site with a huge collection of herbal resources and information. Howie Brounstein is the owner and operator of Columbines and Wizardry Herbs, Inc., of Eugene, Oregon. He teaches courses in botany, herbalism, and ethical wildcrafting. Of particular note are the extensive files on herbal wildcrafting.

Institute for Nutritional Advancement (INA)
<http://www.nutraceuticalinstitute.com/>

The INA is a noncorporate division of Denver-based Industrial Laboratories, an independent laboratory that provides analytical and consulting services to the natural products industry. One of its initiatives is the Methods Validation Program (MVP), an international effort to validate and make available analytical methods to meet the demand for global consistency in the testing of botanicals. INA works with the USP and the NF.

Michael Moore's Herbal-Medical Glossary
<http://chili.rt66.com/hrbmoore/ManualsMM/
MedHerbGloss2.txt>

This useful glossary of medical and botanical terms from the perspective of herbal practitioners is compiled by Michael Moore, director of the Southwest School of Botanical Medicine. (See the related entries in Chapters 2 and 3 for more details on the resources available at the Southwest School site.)

A Mini-Course in MEDICAL BOTANY
<http://www.ars-grin.gov/duke/syllabus/>

These documents constitute the syllabus for a continuing education course taught by the respected ethnobotanist Dr. James A. Duke (see the entry for Dr. Duke's Web site in Chapter 12). They are provided by the Agricultural Research Service (ARS), U.S. Department of Agriculture (USDA). Though ostensibly just a "syllabus," these documents contain valuable information on all aspects of herbs that is not readily available elsewhere. The nineteen modules cover introductory botany, herb formulations, dangerous herbs, Chinese herbs, and medicinal herb use in non-Western pharmacies. Whereas some modules contain only an outline of the material to be covered in the course, others, such as the HDR: Herbal Desk Reference, Formulations, and Last and Least . . . Dangerous Herbs, contain text for the corresponding lecture and are thus more informative. The HDR module contains detailed information on the best-selling Western and Chinese herbs in the United States, with information on activity, indications, adverse effects, contraindications, and dosage. This resource is particularly useful for its lists of African, American Indian, Hawaiian, Arabic, and Ayurvedic (Indian subcontinent) herbs.

Natural HealthLine
  <http://www.naturalhealthvillage.com/>

This is an online newsletter and very useful site providing up-to-date summaries of news articles related to complementary and alternative medicine/health care, including legislative activity and research reports. Many of the articles are authored by Peter Chowka, a veteran medical journalist who has written extensively about both allopathic and complementary and alternative medicine (CAM). The items on Natural HealthLine are provided by Natural HealthLine and Natural Health Village, affiliated with and produced by a group called Project Cure, in Fairfax, Virginia. Links are also provided to selected online news articles.

Chapter 5

Herbal Associations and Organizations

Be it ordained, established and enacted by the authority of this present parliament, that at all time from henceforth, it shall be lawful to every person being the King's subject, having knowledge and experience of the nature of Herbs, Roots and Waters . . . to practise, use and minister in and to any outward Swelling or Disease any Herb or Herbs, Ointments, Baths, Pulters and Emplaisters, according to their Cunning, Experience and Knowledge . . . Without Suit, Vexation, Trouble, Penalty or Loss of their Goods.

The "Herbalists' Charter," 1543

HERBAL ORGANIZATIONS IN THE UNITED STATES AND EUROPE

In the United States, few institutions or organizations offer training in herbal medicine and no nationwide professional body promotes its use. The situation is a little better in several European countries where herbal preparations have a much longer history of use, such as in England and France, or are more effectively integrated into mainstream medicine, such as in Germany. The current explosion of interest in herbs, though, has led to the launching of several new professional organizations in both North America and several European countries. Many of these are not just concerned with offering educational opportunities for herbal practitioners but are seeking to improve the standing of medical herbalists by promoting professional standards and educational programs. Current issues of concern to traditional herbalists in both the United States and Europe include proposed legislation that seeks to restrict the practice of herbal professionals, and the question of how to embrace scientific advancements while being careful not to forget the traditional roots of herbal practice.

The National Institute of Medical Herbalists (NIMH) is the oldest association of practicing herbalists in the Western hemisphere. It was founded in Great Britain in 1864 to promote herbal medicine and to protect the status and rights of the medical herbalist. This was partially a response to legislation introduced into the British Parliament in 1858, part of which was viewed by herbalists as an attempt to prevent the practice of medicine by anyone who had not been trained in a conventional allopathic medical school. Conflicts with Western orthodox medicine have a long tradition, both in the United States and in European countries. As early as the time of the English Tudor King Henry VIII, confrontations occurred between medical doctors and herbalists. During Henry's reign, a law was passed in England titled An Act That Persons, Being No Common Surgeons, May Administer Outward Medicines; this established the rights of herbalists to practice their profession and protected said practitioners from interference by the medical authorities of the day. This is often referred to as the "Herbalists' Charter."

The continued growth and evolution of the European Union (EU) as a unified political entity has important ramifications for the practice of herbal medicine in its individual countries. As the popularity of herbs and dietary supplements continues to grow, Europe is moving toward new legislation that attempts to harmonize the regulation of traditional herbal products.[1] The European Herbal Practitioners Association (EHPA) was founded in 1993 as a result of fears that new EU regulations might adversely affect the availability of herbal preparations in the individual countries. A central aim of the EHPA and other European herbal groups is "to encourage the creation of appropriate European legislation that ensures the continuing right of professional herbal practitioners to access traditional herbal medicines" (EHPA web site).

WEB SITES OF PROFESSIONAL HERBAL ORGANIZATIONS

Alternative Medicine Foundation, Inc.
<http://www.amfoundation.org/>

The Alternative Medicine Foundation, located in Bethesda, Maryland, is a nonprofit organization committed to providing health practitioners and consumers with reliable information about the safety and effectiveness of alternative medicine therapies. It publishes the *Journal of Alternative & Complementary Medicine,* a peer-reviewed journal, and maintains the unique HerbMed database that summarizes evidence on the efficacy of herbal therapies (see the HerbMed entry in Chapter 9).

American Association of Drugless Practitioners (AADP) <http://www.aadp.net/>

The AADP is dedicated to promoting and enhancing the professional image and prestige of Holistic Medical Practitioners, a group that includes herbalists who eschew the use of conventional drugs.

The American Association of Naturopathic Physicians (AANP) <http://www.naturopathic.org/>

The word naturopathy was first coined in the United States around 100 years ago and refers to a holistic way of treating illness using the "natural healing forces" present in the human body. According to the philosophical basis of naturopathy, healing occurs naturally in the human body if it is given what it needs, such as a proper diet, pure water, fresh air, sunlight, exercise, and rest. As with other holistic healing systems, the emphasis is not on finding a disease and curing it, but rather on helping the body establish what is considered to be its own state of good health. Traditional naturopaths do not diagnose or treat disease but instead focus on health and education. Naturopaths employ a wide range of herbal medicines, from a variety of herbal traditions.

A licensed naturopathic physician (ND) attends a four-year, graduate-level naturopathic medical school and is educated in all of the same basic sciences as an allopathic physician. In addition to a standard medical curriculum, the naturopathic physician is required to complete training in clinical nutrition, acupuncture, homeopathic medicine, botanical medicine, and psychology. In the United States, naturopathic physicians are not yet licensed in all fifty states. Membership in the AANP is available only to a naturopathic physician who is a graduate of a recognized college of naturopathic medicine.

This Web site provides a database of articles on naturopathy written for the public, including papers on the adverse effects of botanicals and their place in the treatment of cancer, arthritis, and cardiovascular disorders.

American Botanical Council (ABC)
<http://www.herbalgram.org/>

The American Botanical Council (ABC) is a nonprofit organization whose mission is to educate the public about beneficial herbs and plants and to promote their safe and effective use. It is probably one of the best-known and most active herbal organizations in the United States. The ABC conducts continuing education (CE) modules for pharmacists and, along with the Herb Research Foundation (HRF), publishes the widely respected peer-reviewed journal *HerbalGram*. The ABC site has useful information on upcoming congresses and meetings of relevance to herbal medicine, as well as useful links to some very specialized herbal resources.

HerbalGram provides a wide spectrum of information on the medicinal uses of herbs, including legal or regulatory information, media coverage on herbs, articles on herbal research, market reports, book reviews, and in-depth reviews of specific herbs. It is indexed by the AGRICOLA and IBIDS databases (see the individual database entries in Chapter 13).

In the summer of 1998, the ABC published the English translation of Germany's Commission E Monographs, widely regarded as the most accurate source of information on the safety and efficacy of herbs and phytomedicines (i.e., herb extracts).[2] About 300 monographs have appeared so far, covering most of the economically important herbal remedies available in Germany. Each monograph provides such information as description of the herb, formulation, pharmacological properties, clinical data, recommended doses, and approved uses. To view particular monographs, select Publications at the home page, then follow the Commission E link.

The American Herbal Products Association (AHPA)
<http://www.ahpa.org/>

The American Herbal Products Association (AHPA) is a national trade association representing the U.S. herbal products industry (for more information, see the corresponding entry in Chapter 4).

The American Herbalists Guild (AHG)
<http://www.americanherbalistsguild.com>

Launched in 1989, the American Herbalists Guild (AHG) is a nonprofit educational organization devoted to the promotion of educational excellence, ethical standards, and integrity in herbalism. It is the only peer-reviewed organization for professional herbalists. Useful AHG publications include a directory of educational programs for herbalists, a list of recommended readings, and various position papers on herbal issues. Its *Journal of the American Herbalists Guild* publishes "scholarly manuscripts on all aspects of herbalism with an emphasis on clinical and professional application of herbal products within the vitalistic paradigm." Most of the documents themselves are not online but may be purchased from the AHG for a small fee.

American Society for Pharmacology and Experimental Therapeutics (ASPET)
  <http://www.faseb.org/aspet/>

ASPET Herbal Medicine and Medicinal Plant Interest Group
  <http://www.faseb.org/aspet/H&MIG7.htm#top>

Membership in ASPET is restricted to qualified investigators who have "conducted and published a meritorious original investigation in pharmacology." Individuals holding doctoral degrees (PhD, MD, or equivalent) are considered to be qualified investigators. (Exceptions may be made for those who do not meet the degree requirement but have made major original research contributions to pharmacology.)

ASPET publishes *Drug Metabolism and Disposition, Pharmacological Reviews, Molecular Pharmacology,* and *The Journal of Pharmacology and Experimental Therapeutics*.

ASPET hosts an herbal medicine and medicinal plant special interest group, providing a forum for clinical pharmacologists to discuss phytomedicinals, their active principles, modes of action, and regulatory issues concerning their use. The site has useful book reviews, information on relevant scientific meetings, and updates on the regulatory status of herbal products in the United States.

The American Society of Pharmacognosy (ASP)
  <http://www.phcog.org/>

Pharmacognosy is the study of the physical, chemical, biochemical, and biological properties of drugs, drug substances, or potential drugs of natural origin, as well as the search for new drugs from natural sources. Considered for many years to be one of the least important areas of pharmacy, the resurgence of interest in herbal preparations has led to a growing interest in this discipline. The ASP site lists all degree-granting graduate programs in pharmacognosy or natural products chemistry; it also features a useful list of the major pharmacognosy journals, with links to the respective publishers' Web sites.

Association of Natural Medicine Pharmacists (ANMP)
<http://www.anmp.org/>

The Association of Natural Medicine Pharmacists (ANMP) is a professional association for pharmacists and those interested in the field of "natural medicines," including phytomedicinals, nutrition, homeopathy, and aromatherapy. This site provides useful information on relevant books, monographs, and CD-ROM products. The ANMP has a quarterly newsletter, *The Source,* and although subscription requires AMMP membership, selected articles are available from this site.

The British Herbal Medicine Association
<http://www.ex.ac.uk/phytonet/bhma.html>

The BHMA was founded in 1964 to advance the science and practice of herbal medicine in the United Kingdom. Members of the BHMA include companies involved in the manufacture of herbal medicines or in the supply of botanical drugs, herbal practitioners, academics, pharmacists, and others. The BHMA publishes the *British Herbal Pharmacopoeia* (BHP), a noted resource for assessing the identity and purity of herbal ingredients.[3]

British Herbal Practitioners Association (BHPA)
<http://www.btinternet.com/~nimh/bhpa.html#aims>

The British Herbal Practitioners Association (BHPA) seeks to foster unity within the herbal profession, to promote the availability of professional herbal treatment, and to raise the standards of training and practice within the herbal profession in the United Kingdom. Its sister organization is the European Herbal Practitioners Association (EHPA) (see the corresponding entry in this chapter). Members of the BHPA include the National Institute of Medical Herbalists, the General Council and Register of Consultant Herbalists, the Register of

Chinese Herbal Medicine, the Association of Master Herbalists, and the Association of Traditional Chinese Medicine (UK).

DIRLINE: Directory of Health Organizations
<http://dirline.nlm.nih.gov/>

DIRLINE (Directory of Information Resources Online) is produced by the National Library of Medicine (NLM) and indexes U.S. (with some international) health and biomedical organizations, government agencies, information centers, professional societies, voluntary associations, support groups, academic and research institutions, and research facilities and resources. Each record provides name, address(es), phone number(s), description of services offered, list of publications produced by the group, and any holdings information. DIRLINE contains approximately 14,000 records and focuses primarily on health and biomedicine. Search the directory by keyword or medical subject heading (MeSH) (for appropriate MeSH terms, see the entry for PubMed in Chapter 13).

European Herbal Practitioners Association (EHPA)
<http://www.users.globalnet.co.uk/~ehpa>

The EHPA represents professional herbal practitioners in the European Union (EU). Its sister organization is the British Herbal Practitioners Association (BHPA) (see Chapter 4's entry for more information on the EHPA).

Herb Research Foundation (HRF)
<http://www.herbs.org/>

The Herb Research Foundation (HRF) is a nonprofit research and educational organization focusing on herbs and medicinal plants. Along with the American Botanical Council, it publishes *HerbalGram,* a peer-reviewed journal with a scientific focus on medicinal herbs. This is an invaluable site for keeping up to date with advances in the field of herbal medicine and with the latest results from clinical trials involving herbal products. From the home page, select News Views to view updated news items from around the world, covering herb research and related regulatory issues. Follow the link to Research Reviews to view valuable information on international clinical studies involving herbal preparations. The HRF maintains its own extensive library of herbal resources, and in-depth information profiles on more than 1,000 herbs are available for a fee.

The Herb Society of America
<http://www.herbsociety.org/>

The Herb Society of America focuses on the cultivation of herbs and the study of their history and uses, both past and present. It was founded in 1933 for the purposes of furthering knowledge and use of herbs and contributing the results of the experience and research of its members to the records of horticulture, science, literature, history, the arts, and economics. Membership is open to the general public; for information about member categories and dues, see the Web site. Materials may be requested via e-mail, phone, letter, or in person. Regional and national symposia are conducted by members or units of the society throughout the United States and Canada. The society's Web site provides access to a range of useful resources, including herbal fact sheets (click on Information Resources) and information on educational activities for the general public.

Herb Society (UK) Online
<http://www.herbsociety.co.uk/>

The Herb Society is an organization based in London, England, devoted to the dissemination of information about herbs. It publishes *Herbs,* a well-known consumer-oriented herbal magazine. The Web site provides access to several unique articles on herbal medicine, including *"Cuba: Plants and Medicine"* in Cuba and *"Herb Gardens of Provence."* The section Herbal Legislation monitors changes in herbal legislation in the United Kingdom and the European Union (EU).

Institute for Natural Products Research (INPR)
<http://www.naturalproducts.org/>

The Institute for Natural Products Research (INPR) is a nonprofit educational and scientific organization formed to promote research into botanical medicines and other natural medicines.

**The Institute of Economic Botany
(New York Botanical Garden)
  <http://websun.nybg.org/bsci/ieb/>**

The field of economic botany involves the study of the relationship between plants and people. The New York Botanical Garden (NYBG) Institute of Economic Botany (IEB) was founded in 1981 to focus some of the NYBG's research activities on "applied botanical questions of human concern." Current activities include projects involving Belize, Bolivia, Brazil, Colombia, Dominica, Dominican Republic, Ecuador, French Guiana, Guatemala, Guyana, Honduras, Indonesia, Martinique, Panama, Paraguay, Peru, Thailand, the United States, and the Virgin Islands. This work includes the identification of plants used by local people for food, fiber, fuel, and medicine, and their evaluation by the National Cancer Institute for use as anti-AIDS and anticancer therapies.

**Integrative Medicine Institute (IMI)
  <http://integrativemed.org/>**

The IMI conducts research into the effectiveness, safety, patient satisfaction, and cost of integrating complementary and alternative therapies with conventional medicine. Research areas include chiropractic, botanical medicine, homeopathy, massage, nutritional medicine, oriental medicine, osteopathic medicine, and therapeutic yoga.

**National Herbalists Association of Australia (NHAA)
  <http://www.nhaa.org.au/>**

The NHAA was founded in 1920 and is Australia's only professional herbal association. It publishes the peer-reviewed *Australian Journal of Medical Herbalism*. Of particular interest is the Native Herb Forum Web page, which seeks to "positively influence the clinical use of Australian herbs; and ensure ongoing supplies by exploring their potential for cultivation." The Forum Web site houses review articles on Australian medicinal plants and fact sheets and monographs on native Australian plants, such as sneezeweed *(Centipeda minima)* and kangaroo apple *(Solanum aviculare)*.

The National Institute of Medical Herbalists (NIMH)
<http://www.btinternet.com/~nimh/>

The institute was originally established in 1864 as the National Association of Medical Herbalists and is one of the leading professional herbal organizations in the United Kingdom. The Web site provides some introductory information on herbs and herbal medicine, including an FAQ section, information on education, and an overview of research in this area. The institute publishes the *European Journal of Herbal Medicine* and *Greenfiles,* a quarterly newsletter of research abstracts. The site also features a useful section on herbal teas, including their preparation and use.

The Research Council for Complementary Medicine (RCCM)
<http://www.rccm.org.uk/>

The RCCM, based in the United Kingdom, carries out, promotes, and evaluates rigorous research in complementary medicine to encourage safe, effective practice and improved patient care. The RCCM produces a bibliographic database, Centralised Information Service in Complementary Medicine (CISCM), that currently has information on over 4,000 randomized trials and over 60,000 citations and abstracts, covering all the major complementary therapies (for more information on the CISCM database, see the corresponding entry in Chapter 13).

Society for Medicinal Plant Research
(Gesellschaft für Arzneipflanzenforschung)
<http://www.uni-duesseldorf.de/WWW/GA/>

Based in Germany, but with an international reach, the Society for Medicinal Plant Research is one of the leading European organizations devoted to the advancement of research in the field of medicinal plants. Areas of interest include the biological and pharmacological activity of natural products, the breeding and cultivation of medicinal plants, and the development of manufacturing and quality control standards for herbal products. The society publishes an online newsletter, available from the Web site, and *Planta Medica,* one of the leading research journals in herbal medicine. Tables of contents, with abstracts, for recent issues of the journal can be viewed for free.

Chapter 6

The Mainstreaming of Herbal Medicine

Between tradition and modernity there is a bridge. When they are mutually isolated, tradition stagnates and modernity vaporizes; when in conjunction, modernity breathes life into tradition, while the latter replies with depth and gravity.

Octavio Paz
"In Search of the Present,"
Nobel Lecture, 1990

UNIVERSITY-AFFILIATED, NONPROFIT, AND OTHER RESEARCH INSTITUTIONS

The previous chapter was primarily devoted to those organizations and associations working within what can be regarded as the traditional herbal movement. In the United States, this movement has largely operated on the fringes of mainstream Western medicine. Historically, allopathic physicians in the United States have vigorously fought CAM practices, denouncing them as quackery and attacking them for being unscientific.[1] (Throughout this book, I prefer to use the term Complementary and Alternative Medicine, or CAM, to refer to what is generally called "alternative" medicine. See the glossary for a definition of CAM and allopathy.) However, whatever the views of allopathic physicians, a sizable section of the American public is turning to CAM practitioners. This has forced mainstream health organizations and medicine to reexamine CAM therapies, assessing their effectiveness and trying to determine how best to incorporate them into the routine practice of medicine. A recent article reported that 64 percent of U.S. medical schools have already begun to offer some basic instruction in CAM medicine practices.[2] Hospitals are beginning to integrate such practices into health care programs, and an increasing number of medical insurance programs are offering benefits packages that include the services of CAM practitioners.[3]

In response to criticism that much of the herbal literature relies too heavily on anecdotal evidence and ancient folktales, physicians and researchers have focused a lot of their recent activity on gathering evidence on the efficacy and safety of herbal treatments. One of the most significant developments in this area was the creation of the Office of Alternative Medicine (OAM) in 1992. Established by congressional mandate as an office of the National Institutes of Health (NIH), and thus funded by the U.S. government, the OAM was set up to promote research into the effectiveness of CAM practices and to provide the American public with reliable information about their safety and effectiveness.[4] In 1998, the office was elevated to an NIH center, with an expanded mandate and the administrative authority to design and manage its own research programs.[5] To reflect these changes, the name changed to the National Center for Complementary and Alternative Medicine (NCCAM).

To accomplish its goals, the NCCAM cooperates with other institutions to conduct clinical trials, as well as funding several CAM research centers outside the NIH. It provides funding to eleven research centers that evaluate CAM treatments, such as the CAM Research Center for Cardiovascular Diseases at the University of Michigan and the Center for CAM Research in Aging and Women's Health at Columbia University. In 1994, it helped establish an AIDS Research Center at Bastyr University, a leading U.S. college of naturopathic medicine. As part of its goal of disseminating reliable information to health professionals and the American public, the NCCAM produces the CAM section of the Combined Health Information Database (CHID) and has partnered with the National Library of Medicine (NLM) to create the CAM on PubMed database. (See the corresponding entries in Chapter 13.)

The 1990s witnessed the creation of several other CAM research centers and programs. The Richard and Hinda Rosenthal Center for Complementary and Alternative Medicine was established by Columbia University's College of Physicians and Surgeons, the first such resource located at an American university medical school. The University of Arizona has a well-known program in integrative medicine, while institutions such as Harvard, Stanford, and Philadephia's Jefferson College have their own CAM programs.

WEB SITES OF HERBAL RESEARCH INSTITUTIONS

Bastyr University
<http://www.bastyr.org/>

Botanical Medicine Department
<http://www.bastyr.edu/academic/botmed/>

Bastyr University is an accredited private institution located in Kenmore, just north of Seattle, Washington. It was founded in 1978 and named after Dr. John Bastyr, a well-known Seattle naturopathic physician and teacher. It is one of only three accredited naturopathic medical schools in the United States (the others are the National College of Naturopathic Medicine in Portland, Oregon, and the Southwest College of Natural Health Sciences in Scottsdale, Arizona). Bastyr offers courses in naturopathy, homeopathy, acupuncture, vitamin and herbal treatments, chiropractic studies, and meditation; it also operates a clinic in Seattle. In 1994, a Bastyr University AIDS Research Center (BUARC) was established, under a cooperative grant from the NIH's National Institute of Allergy and Infectious Disease (NIAID) and the National Center for Complementary and Alternative Medicine (NCCAM).

Naturopathic medicine employs a wide range of herbal medicines, from a variety of herbal traditions. Go to Bastyr's Department of Botanical Medicine site for general information and resources, including fact sheets on such herbs as stinging nettle *(Urtica dioica),* dandelion *(Taraxacum officinalis),* burdock *(Arctium lappa),* and marshmallow *(Althea officinalis)*. The university now offers Web-based courses on herbal medicine and other CAM areas through its distance learning program.

Center for Alternative Medicine Research and Education (CAMRE) (Harvard Medical School)
<http://www.bidmc.harvard.edu/medicine/camr/>

The Center for Alternative Medicine Research and Education is located in Boston, at the Beth Israel Deaconess Medical Center (Harvard Medical School). Its director, David Eisenberg, MD, is a well-known advocate for the investigation of CAM therapies. His 1993 research study "Unconventional Medicine in the United States: Prevalence,

Costs and Patterns of Use" helped initiate a major reassessment of the use of complementary and alternative remedies in the United States.[6] CAMRE provides opportunities for clinical and teaching experiences in CAM, sponsors several education programs, and helps organize conferences, such as the International Scientific Conference on Complementary, Alternative, and Integrative Medicine Research.

Center for Alternative Medicine Research in Cancer: University of Texas (UT-CAM)
<http://www.sph.uth.tmc.edu/utcam/>

The University of Texas Center for Alternative Medicine Research, housed at the University of Texas—Houston School of Public Health, specializes in the study of CAM therapies for cancer prevention and control. The center is one of thirteen such centers funded by the National Center for Complementary and Alternative Medicine (NCCAM). Go to the Review of Therapies section for fact sheets on herbs with possible anticancer properties, including detailed reviews and analyses of the scientific studies. Information is also available for some of the more esoteric herbs and herbal mixtures, such as cat's claw *(Uncaria tomentosa),* essiac (reputed to be derived from an Indian herbal formula), and the Hoxley formula (a mixture of herbs).

(At the time of writing, the status of the UT-CAM center and its Web site is unclear and information has not been updated since 1998. However, fact sheets and other documents are still accessible and links are operational. The site is in the process of being redesigned by staff at the M.D. Anderson Cancer Center who are currently updating the material with plans to release the new version in the fall of 2001.)

Center for Complementary & Alternative Medicine Research in Asthma
<http://www-camra.ucdavis.edu/>

This site is maintained by the University of California, Davis, through a cooperative agreement with the National Center for Complementary and Alternative Medicine (NCCAM). The Center promotes research into the use of alternative therapies for the treatment of asthma. At the home page, click on Useful Complementary and Alternative Medicine Information & Links for information related to complementary and alternative medicine therapies for asthma, including information about relevant research interests at the University of California, Davis.

The Center for Integrative Medicine at Thomas Jefferson University Hospital
<http://www.jeffersonhospital.org/>

To access, select Centers and Programs from the home page, then Alternative Medicine—Center for Integrative.

This center is affiliated with Thomas Jefferson University Hospital, in Philadelphia, and promotes research and education in different CAM fields through "integrative medicine." (See this chapter's entry for University of Arizona Program in Integrative Medicine.) The Web site provides access to well-written guides on various main areas of CAM, including herbs. Of special interest is the center's examination of Latin American community-based practices, such as the 500-year-old tradition of Chicano folk medicine known as "curanderismo," a healing system popular among large Mexican-American communities in the U.S. Southwest, employing rituals, potions, and herbal remedies.

Complementary and Alternative Medicine Program at Stanford University (CAMPS)
<http://camps.stanford.edu>

CAMPS conducts research into "successful aging" and the specific effects of selected complementary and alternative medicine therapies that could affect this process. The concept of "successful aging" implies that an individual remains mentally and physically vital and productive well into old age. The program's mission is "to study the effects of complementary and alternative medicine (CAM) therapies that may enhance successful aging, reduce frailty, and increase independence and quality of life in older persons, especially those therapies that may decrease the impact of cardiovascular and musculoskeletal diseases." Research projects include clinical studies of the effects of ginkgo biloba and the Ayurvedic herb gotu kola *(Centella asiatica)* on cognitive function.

Complementary Health Studies, University of Exeter (United Kingdom)
<http://www.ex.ac.uk/chs/>

Complementary Health Studies, at the University of Exeter, United Kingdom, was created in 1987 by Simon Mills, a medical herbalist, and Roger Hill, a professional acupuncturist. It provides advanced teaching and research opportunities for health care professionals in complementary health studies. Research interests include the quality, safety, and efficacy of herbal treatments and the organization of complementary and alternative medicine in the United Kingdom. The center compiled and produced the new and enlarged edition of the *British Herbal Pharmacopoeia,* published in 1996. One of its most important initiatives is PhytoNET, a Web site collecting information on herbal adverse reactions (see the PhytoNET entry in Chapter 10).

Fogarty International Center (FIC)
<http://www.nih.gov/fic/>

Named after Congressman John E. Fogarty, the FIC is part of the U.S. National Institutes of Health (NIH). Its mission is "to reduce global health disparities by supporting and promoting research and to prepare the current and future generation of international and U.S. scientists to meet global health needs." FIC supports basic biological, behavioral, and social science research, as well as related research training. One of the center's areas of concern is the consequences for human health of the loss of the planet's biodiversity: its International Cooperative Biodiversity Groups (ICBG) program is a unique effort that addresses how local communities can benefit economically from the discovery and conservation of natural drug products. Reports prepared by the Fogarty Center address the significance and use of both prescription and herbal remedies, including the latter's place in the health care systems of indigenous peoples.

Medicinal Plant Research Centre (U.K.)
<http://www.newcastle.ac.uk/medplant/>

The Medicinal Plant Research Centre is affiliated with the University of Newcastle Upon Tyne, in the north of England. Established in 1996, its mission is to investigate plant species for new drugs. Research is focused on the treatment of neurological disorders, involving herbs such as sage *(Salvia officinalis)* and lemon balm *(Melissa officinalis)* that have been found to contain compounds that show promise in alleviating the symptoms of Alzheimer's disease. A trial of lemon balm aromatherapy is also in progress. From the home page go to the Recent Publications section to see details of the Center's research interests.

National Center for Complementary and Alternative Medicine (NCCAM)
<http://altmed.od.nih.gov/nccam/>

NCCAM identifies and evaluates "unconventional" health care practices and supports research projects to assess their effectiveness. Its goal is to incorporate those complementary and alternative medical practices that have been proven safe and effective into mainstream medicine. Congress originally created an Office of Alternative Medicine (OAM) in 1991, or, more accurately, it was created by Senator Tom Harkin, the chairman of the Senate Appropriations Subcommittee with jurisdiction over funding for the NIH. The OAM had a mandate to evaluate research into alternative treatments and to serve as an information clearinghouse for health professionals and consumers. At the time, many critics denounced the new office as a waste of taxpayer money, arguing also that it would open the door for fraud and quackery.[7] However, the center continues to garner support and, in 1998, was reorganized and elevated to a full-fledged center within the NIH, with an increased budget and the authority to conduct or support research and investigations, epidemiological studies, basic science research, as well as clinical trials.

As might be expected, the NCCAM Web site has plenty of useful information aimed at both health professionals and consumers. To navigate this site, use the Site Map. Resources include general articles on complementary and alternative medicine practices, notices of upcoming international meetings, opportunities for NIH research and research funding, and fact sheets and FAQs on CAM and specific herbs, such as St. John's wort. Also included are news items documenting the activities of NCCAM, as well as updates on NCCAM-funded clinical trials.

Sometimes confusion or disagreement arises over what is, and what is not, complementary and alternative medicine, and how they should be classified and described. NCCAM groups CAM practices within the following five major "domains":

1. Alternative medical systems (Ayurvedic medicine, traditional Chinese medicine)
2. Mind-body interventions (music, art, prayer)
3. Biologically based treatments (herbal, dietary)
4. Manipulative and body-based methods (chiropractic, massage)
5. Energy therapies (qi gong, magnetic therapy, therapeutic touch)

One of the goals of the NCCAM is to "distribute scientifically based information about CAM research, practices, and findings to health care providers and consumers." The center has therefore initiated several novel projects to improve and increase dissemination and use of the CAM journal literature. Of these, one of the most important is "CAM on PubMed," a subset of the PubMed database. The center also selects resources for the more consumer-oriented Combined Health Information Database (CHID) (for detailed information on both of these databases, see the corresponding entries in Chapter 13).

National Institute of Allergy and Infectious Diseases (NIAID) <http://www.niaid.nih.gov/>

The National Institute of Allergy and Infectious Diseases (NIAID) is the major federal institute for supporting research into infectious, immunologic, and allergic diseases. NIAID funds a number of centers for tropical medicine research in countries where diseases such as malaria, filariasis, trypanosomiasis, and leprosy are endemic. In collaboration with Johns Hopkins University and the University of Miami School of Medicine, NIAID supports studies on "qinghaosu," an antimalaria drug used in traditional Chinese medicine. Qinghaosu is extracted from the sweet wormwood plant *Artemisia annua*.

New Center for CAM Research in Aging and Women's Health <http://cpmcnet.columbia.edu/dept/rosenthal/Aging.html>

The New Center for CAM Research in Aging and Women's Health is located at the Richard and Hinda Rosenthal Center for Complementary and Alternative Medicine, Columbia University (see the corresponding entry in this chapter). It is one of several research centers funded by the NCCAM. The center provides information about the short- and long-term benefits of complementary and alternative therapies, especially herbs and the effects of diet. Research addresses important questions regarding the modulation of hormonal function and its impact on the health of women as they age. Other areas of interest include cardiovascular health, cancer, and osteoporosis, as well as the relief of symptoms associated with menopause. One study is currently looking at the effect of black cohosh *(Cimicifuga racemosa)* on menopausal hot flashes.

Oregon Center for Complementary and Alternative Medicine in Neurological Disorders (ORCCAMIND)
<http://www.ohsu.edu/orccamind/>

Oregon Health Sciences University (OHSU) has joined forces with five other Oregon institutions to form the Oregon Center for Complementary and Alternative Medicine in Neurological Disorders (ORCCAMIND). Member institutions include the National College of Naturopathic Medicine, Western States Chiropractic College, the Oregon College of Oriental Medicine, the Linus Pauling Institute at Oregon State University, and the Portland Veteran Affairs Medical Center. ORCCAMIND is based at OHSU. In 1999, the center was awarded a $7.8 million grant from the National Institutes of Health to identify and study CAM, focusing on such neurological disorders as Alzheimer's disease and multiple sclerosis. Specific projects include research into the use of CAM antioxidants to treat neurodegenerative (Alzheimer's and Parkinson's) and demyelinating diseases (multiple sclerosis), and studies on interactions between herbal supplements and over-the-counter (OTC) medications.

The Osher Center for Integrative Medicine (OCIM)
<http://www.ucsf.edu/ocim/>

The OCIM is named after Bernard Osher, a businessman and philanthropist from the San Francisco Bay area of California. Affiliated with the School of Medicine at the University of California, San Francisco, the center was established in 1997 to investigate "integrative medicine" and "to search for the most effective treatments for patients by combining nontraditional and traditional approaches that address all aspects of health and wellness." Current areas of interest include a Tibetan medicine project to assess the efficacy of Tibetan herbs in treating advanced breast cancer (see the entries Tibetan Medicine Bibliography and Tibetan Plateau Project in Chapter 7) and studies involving the use of saw palmetto *(Serenoa repens)* extract to treat benign prostatic hyperplasia (BPH).

Pediatric Center for Complementary and Alternative Medicine at the University of Arizona College of Medicine (PCCAM)
<http://www.peds.arizona.edu/>

Steele Memorial Children's Research Center
<http://www.crc.arizona.edu/research/alt-therapy.htm>

Situated at the University of Arizona in Tucson, the PCCAM is one of thirteen centers funded by the National Center for Complementary and Alternative Medicine (NCCAM). It is the first center for CAM research on pediatric diseases to be established in the United States. Studies focus on "pediatric disorders for which no conventional therapy is available." Current projects include a randomized, controlled trial of osteopathy and echinacea for recurrent ear infections and an evaluation of relaxation, guided imagery, and chamomile tea for functional abdominal pain. The center hopes to establish a pediatric research fellowship in CAM research methodologies.

The Richard and Hinda Rosenthal Center for Complementary and Alternative Medicine
<http://cpmcnet.columbia.edu/dept/rosenthal/>

The Richard and Hinda Rosenthal Center for Complementary and Alternative Medicine was established in 1993 by Columbia University's College of Physicians and Surgeons. In addition to promoting and conducting research into alternative and complementary therapies, it provides comprehensive information on alternative and complementary health care practices. The center's Web site provides information on a wide range of resources for researchers, health care practitioners, and the public. Physicians and medical students will find valuable information about residency and fellowship programs in alternative and complementary medicine and education and training programs in herbal medicine.

The Rosenthal Center was awarded a grant from the National Center for Complementary and Alternative Medicine (NCCAM) to develop a New Center for CAM Research in Aging and Women's Health (see the corresponding entry in this chapter). It has also established a Cancer Information Center to make available reliable information on CAM cancer therapies.

Rocky Mountain Center for Botanical Studies, Inc.
<http://www.herbschool.com/>

The Rocky Mountain Center for Botanical Studies (RMCBS) offers educational programs in Western herbalism, focusing on indigenous North American plants. The school is approved and regulated by the state of Colorado, Division of Private Occupational Schools, and is an associate member of the American Herbalists Guild (AHG) and the American Herb Association (AHA). It offers advanced herbal studies program and a clinical herbalism internship, which is available to advanced students of herbalism.

Rocky Mountain Herbal Institute (RMHI)
<http://www.rmhiherbal.org/>

RMHI TCM Herb Library
<http://www.rmhiherbal.org/ai/pharintro.html>

Located in Hot Springs, Montana, the Rocky Mountain Herbal Institute (RMHI) is a private educational and research organization, offering professional training in Chinese herbal sciences. Its herbal sciences program has been accredited by the American Association of Drugless Practitioners (see the corresponding entry in Chapter 5). The Web site provides a wealth of information on Chinese herbal science and related health topics, as well as online access to chapters from its *Textbook of Traditional Chinese Herbal Sciences*. Also of note is the RMHI Herb Library, with information on over 200 herbs in the traditional Chinese medicine materia medica. Entries provide information about botanical classification, dosage, contraindications, preparation, physiological notes, and traditional Chinese clinical indications. Searching the database is free, but registration is required: to access, select HerbLibrary from the home page.

UIC/NIH Center for Botanical Dietary Supplements Research (University of Illinois at Chicago College of Pharmacy)
<http://www.uic.edu/pharmacy/research/diet/>

The University of Illinois at Chicago College of Pharmacy is one of the World Health Organization (WHO)'s two collaboration centers in the United States. It produces the NAPRALERT (NAtural PRoducts ALERT) database, containing bibliographic and factual data on natural products, such as plant, microbial, and animal (including marine) extracts (see the NAPRALERT entry in Chapter 13).

The UIC/NIH Center for Botanical Dietary Supplement Research is a new initiative, designed to employ a multidisciplinary approach in studies involving the clinical safety and efficacy of botanicals. One initial focus will be on herbs used to treat women's health, such as black cohosh *(Cimicifuga racemosa)* and red clover *(Trifolium pratense),* which show promise in the relief of symptoms associated with menopause.

University of Arizona Program in Integrative Medicine
<http://integrativemedicine.arizona.edu/>

Many medical institutions in the United States, including medical schools, are beginning to approach CAM through what is known as "integrative medicine." Integrative medicine seeks to combine the best of both traditional Western medicine and nontraditional alternative medicine approaches and techniques to address the biological, psychological, social, and spiritual aspects of health and illness.

The director of the program at Arizona is Andrew Weil, MD, an internationally known author of books and many scientific and popular articles on the incorporation of CAM interventions into modern medical practice. Dr. Weil also has a comprehensive Web site intended to answer consumer questions (see the Ask Dr. Weil entry in Chapter 12).

University of Maryland Complementary Medicine Program (CMP)
<http://www.compmed.ummc.umaryland.edu/>

The Center for Alternative Medicine Evaluation and Research in Arthritis (CAMERA)

The Complementary Medicine Program (CMP) at the University of Maryland School of Medicine was founded in 1991. Research focuses on the investigation of the effectiveness of complementary therapies for the alleviation of conditions accompanied by acute and chronic pain, such as arthritis. CAM areas of interest include herbal medicine, homeopathy, and acupuncture.

In 1999, the National Center for Complementary and Alternative Medicine (NCCAM) awarded a $7.8 million grant to the University of Maryland to establish a center for alternative medicine research in arthritis and related disorders. The Center for Alternative Medicine Evaluation and Research in Arthritis (CAMERA) is housed within the Complementary Medicine Program (CMP). Though the focus is currently on investigating the use of acupuncture, future projects involve studies with botanicals, including traditional Chinese herbal formulas.

CAMERA produces the Complementary and Alternative Medicine Pain Database (CAMPAIN) (see the corresponding entry in Chapter 13).

The University of Michigan Complementary and Alternative Medicine Research Center for Cardiovascular Diseases (UM-CAMRC)
<http://www.med.umich.edu/camrc/>

The UM-CAMRC was established to analyze and assess complementary and alternative medicine interventions that might prove to be useful in the prevention, management, and treatment of cardiovascular diseases. Current projects include studies on hawthorn *(Crataegus oxycantha)* involving individuals with chronic heart failure. Hawthorn has a well-established history in traditional herbalism as a general "tonic" for the heart and circulatory system.[8]

Chapter 7

Traditional Medicine, Folklore, and Ethnobotany

> . . . everything on the earth has a purpose, every disease an herb to cure it, and every person a mission. This is the Indian theory of existence.
>
> Mourning Dove Salish, 1888-1936

TRADITIONAL MEDICINE AND HERBAL REMEDIES

According to the World Health Organization (WHO), about 4 billion people, or 80 percent of the world's population, use herbal remedies for some aspects of their health care.[1] WHO uses the term *traditional medicine* to refer to "ways of protecting and restoring health that existed before the arrival of modern medicine" (WHO Web site). It encompasses many systems of healing that are not based on the Western scientific approach and are often centered around cultural beliefs and practices handed down from one generation to another. The concept includes mystical and magical rituals, herbal therapy, and other treatments. It can encompass African traditional medicine, Arabic medicine, Ayurvedic medicine, and traditional Chinese medicine. In many developing countries, traditional medicine is still the main source of primary health care, and herbs play a significant role in these healing systems.

In many instances, the efficacy of centuries-old traditional herbal treatments has been confirmed by modern research. For example, extracts of the Ayurvedic herb *Psoralea corylifolia* have been used for over 3,000 years to treat a skin pigmentation condition know as vitiligo.[2] Our grandmothers also seem to have been right to recommend chamomile *(Matricaria recutita)* tea as soothing relief for indigestion, since modern research indicates that it does indeed have anti-inflammatory properties, making it a valuable aid for the relief of several gastrointestinal complaints.[3]

TRADITIONAL CHINESE, TIBETAN, AND AYURVEDIC MEDICINE

China and the Indian subcontinent have developed what are probably the two most widely known systems of traditional health care: traditional Chinese medicine (TCM) and Ayurvedic medicine. TCM refers to an ancient healing system that is thought to predate current Western medicine by over 2,000 years.[4] It encompasses a range of seemingly disparate techniques and materials, including widely popular areas such as acupuncture, massage, and herbal preparations, as well as the more esoteric areas of moxibustion and qi gong. Ginseng *(Panax ginseng)* is probably the most famous medicinal herb used in traditional Chinese medicine, and the one best know to Westerners. Many other herbs, such as ma huang and gingko biloba, have been used in Chinese medicine for over 3,000 years but are only now enjoying great popularity in Western industrialized countries.

Ayurvedic medicine, another ancient system of healing, originating before 2500 B.C., provides an integrated or holistic approach to the prevention and treatment of illness using a combination of diets, drugs, and certain practices.[5] The term *Ayurvedic* has its origins in the Sanskrit words *ayus,* meaning life, and *veda,* meaning knowledge or science. Ayurvedic preparations are complex and their main ingredients are plant extracts. Around 1,250 plants are currently in use. Well-known Ayurvedic plants include the snakeroot plant *(Rauwolfia serpentina),* known in India as Sarpaganda. Sarpaganda has been used for centuries to treat insanity as well as physical illnesses such as fever. An extract was isolated in 1952 and subsequently marketed as the drug reserpine for lowering high blood pressure.[5] Many Western herbs generally require little processing and are normally used by themselves in infusions or in tinctures. By contrast, Chinese and Ayurvedic herbs often require complicated processing and are not generally used singly, but in combination with other herbs to form compounds.

Specific mention should also be made here of Tibetan medicine, a healing system that combines knowledge and traditions of Chinese and Indian medicine with a strong spiritual component derived from Buddhism.[4] Though the Tibetan materia medica includes animal and mineral ingredients, it is mainly herbal, relying on some 1,000 plants. Tibetan medical plants include pomegranate *(Punica granatum),* fennel *(Foeniculum vulgare),* and white sandalwood *(Santalum album).* Several traditional Tibetan herbal preparations show great promise as therapeutic agents, such as "Padma-28," currently being studied by several groups for its anti-inflammatory activities.[6,7] This is a mixture of twenty-two herbs, including spiral flag, Iceland moss, chinaberry, myrobalan, cardamom, red saunders, sorrel, camphor, hardy orange, columbine, licorice, ribwort, knotgrass, golden cinquefoil, clove, gingerlily, heart-leaved sida, lettuce, valerian, and marigold.[7]

ETHNOBOTANY

With the renewed Western interest in examining plants used for centuries in various traditional healing systems, the field of ethnobotany has emerged. Ethnobotany involves studies of the use of indigenous plants by people of various cultures in different parts of the world. In the United States, James Duke is one of the best-known ethnobotanists and a widely respected authority and author on medicinal herbs and herbal healing traditions; he is the author of *The Green Pharmacy* and *CRC Handbook of Medicinal Herbs.*[8,9] His ethnobotanical database is a unique source of information on the use of plants by various cultures around the world.

HERBAL FOLKLORE

It should be remembered that herbs generally have a long history of use, with some dating back several thousand years. For some there are written records, detailing uses and observations, while information about others survived only through oral traditions. In traditional herbalism, this "herbal folklore" has been a rich source of information. The word folklore is often used in the context of traditional medicine, referring to the traditional beliefs, legends, and customs of a particular group of people. In many countries, the tradition of "folk medicine" originated in centuries-old indigenous healing systems that flourished long before the development of Western scientific medicine. Folk medicine usually implies the care of the sick by unlicensed healers, or "folk healers," including those who practice herbal medicine. Herbal folklore has played an important role in preserving knowledge about herbal remedies and, in recent years, has become of increasing interest to many people in the modern scientific medical community.

WEB SITES FOCUSING ON TRADITIONAL HERBAL MEDICINE

AACP Basic Resources for Pharmaceutical Education
<http://www.aacp.org/Resources/Reference/Basic_Resources/00_bklst_intro.html>

This is a list of recommended major print and nonprint resources compiled by the Libraries/Educational Resources Section of the American Association of Colleges of Pharmacy (AACP). It is intended to provide information that will assist with collection development and accreditation as applied to the curriculum of the particular school/college of pharmacy for which it is being used. Go to the Pharmacognosy and Natural Products section for selected books dealing with medicinal herbs from around the world.

Africa Journals Online (AJOL)
<http://www.inasp.org.uk/ajol/>

International Network for the Availability of Scientific Publications (INASP)
<http://www.inasp.org.uk/>

AJOL 2000 offers the tables of contents and abstracts of articles from up to fifty journals in the agricultural sciences, science and technology, as well as health and social sciences, published in African nations. Registration is required but the basic bibliographic information is free. AJOL is produced by the International Network for the Availability of Scientific Publications (INASP), an international cooperative network of partners devoted to promoting universal access to reliable information for health care workers in developing and transitional countries. Journal titles include *African Journal of Reproductive Health* (Nigeria), *Central African Journal of Medicine* (Zimbabwe), *Ethiopian Journal of Health Development,* and the *Journal of Medicine and Medical Sciences* (Nigeria).

AfricaBib
<http://www.africabib.org/>

AfricaBib consists of two bibliographic databases covering the African periodical literature: the Bibliography of Africana Periodical Literature Database and the African Women's Bibliographic. The databases were created and are maintained by Davis Bullwinkle, director of the Institute for Economic Advancement (IEA) Research Library, Little Rock, Arkansas. Titles indexed in these database represent materials from over twenty-two nations within North America, Europe, Africa, and Asia.

Bibliography on Native Americans in Agriculture: Ethnobotany/Traditional Medicine
<http://nalusda.gov/outreach/Medicine.htm>

This is a useful bibliography compiled from the AGRICOLA database by Susan Wilzer and Deborah Richardson at the National Agricultural Library (NAL).

California School of Traditional Hispanic Herbalism
<http://www.hispanicherbs.com/>

The California School of Traditional Hispanic Herbalism is devoted to the teaching and preservation of the Hispanic healing tradition known as "curanderismo," a Chicano traditional healing system.

California Society for Oriental Medicine (CSOM)
<http://www.cyberweb1.com/csom/>

The CSOM is a nonprofit association dedicated to promoting the practice of acupuncture and other forms of "Oriental medicine" in California. This site provides much useful information on herbs used in traditional Chinese medicine (TCM), including information on Chinese herbal patent medicines that contain Western drugs, news items documenting legal and political issues, and information on herbal products that are subject to import restrictions, such as aconite *(Aconitum napellus),* an herb much used in homeopathic medicine.

Chinese Cultural Studies: The Chinese Language and Writing
<http://acc6.its.brooklyn.cuny.edu/~phalsall/texts/chinlng2.html>

This is an introductory article on the Chinese language and the pin yin system of romanizing Chinese characters from a course in Chinese cultural studies, given at Brooklyn College of the City of New York.

An increasing amount of information on Chinese herbs is appearing on the Internet. Some of these sites may organize information by the Chinese name for an herb or an herbal remedy, and the user may be given a choice of accessing entries using a pin yin index or a Latin index. For purposes of identification, it is important to use both the Latin binomial and the Chinese pin yin nomenclature when discussing a Chinese herb, as it identifies the plant part and method of preparation of these complicated herbal products. Pin yin, meaning *spell sound,* is a system for writing Chinese sounds using the Western alphabet, also known as the Latin or Roman alphabet. This Pin yin method was introduced by the People's Republic of China and has been in official use in the West since 1958, replacing the older, less correct Wade-Giles system, hence the change from Peking to Beijing. This system has been the standard for the U.S. Government for more than two decades and is now used by the United Nations and most of the world's media.

Crane Herb Company
<http://www.craneherb.com/>

The Crane Herb Company is a major distributor of Chinese herbs, herbal formulas, and acupuncture supplies. This is a useful site for information on the composition of Chinese herbal formulas and their uses.

"Curanderismo": Chicano Folk Medicine
at Thomas Jefferson University Hospital
<http://www.jeffersonhospital.org/show.asp?durki=5304>

The Center for Integrative Medicine at Thomas Jefferson University Hospital has an active research program in the Chicano folk medicine tradition known as "curanderismo," a healing system popular among large Mexican-American communities in the U.S. Southwest. This tradition of folk healing involves the use of herbs, massage, and rituals combined with elements of spiritualism, psychic healing, and magic.

EthnobotDB—Worldwide Plant Uses
<http://ars-genome.cornell.edu/Botany/aboutethnobotdb.html>

EthnobotDB provides information on the medicinal use of over 80,000 plants distributed throughout the world. It was compiled by Dr. James A. Duke and Stephen M. Beckstrom-Sternberg and is currently hosted by the National Germplasm Resources Laboratory (NGRL) at the USDA. The database may be searched by keyword or browsed via drop-down menus listing common name, country, taxonomic class (i.e., genus, family), or use (e.g., analgesic, ache [ear], appendicitis).

Flora Celtica
<http://www.rbge.org.uk/data/celtica/>

The Flora Celtica project seeks to gather information on the role of plants in the lives of the people of Scotland and Celtic Europe, including Wales, the Isle of Man, Ireland, Cornwall, Brittany, northwest Spain, and northern Portugal.

The Herbal Encyclopedia
<http://www.wic.net/waltzark/herbenc.htm>

This site mixes basic information on popular herbs with their religious or mystical symbolism, reflecting the interests of its creator, Lisa Waltz. (See the corresponding entry in Chapter 4 for details.)

HerbWeb
<http://www.herbweb.com/>

This Web site, based on Tim Johnson's book *CRC Ethnobotany Desk Reference,* is an excellent source of information on plant use by many indigenous cultures.[10] The database includes information on over 29,000 species. Select Indigenous Use to find information on herbs used by the American Navajo Indian tribes, the Japanese Ainu people, and the Mayan Indians of Central America.

The History of Chinese Medicine Web Page
<http://staff.albion.edu/history/chimed/>

This site is the creation of Yi-Li Wu, Department of History, Albion College, Albion, Michigan, and Christopher Cullen, Senior Lecturer in History of Chinese Science and Medicine, Department of History, School of Oriental and African Studies, University of London. It is intended as a clearinghouse for scholars interested in the history of Chinese medicine. It contains useful links for scholars, including the full text of some classic Chinese medical texts and bibliographies of Western-language sources on the Chinese materia medica.

History of Traditional Chinese Medicine
(Karolinska Institutet Library)
<http://www.mic.ki.se/China.html>

The Karolinska Institutet in Stockholm is Sweden's only medical university, with a prestigious international reputation. The Karolinska Institute Library provides consumers, health care professionals, and researchers with a large collection of links to diseases, disorders, and related-topic resources. This is a useful collection of resources on the history of traditional Chinese medicine, including a bibliography of Western-language sources on medicine in East Asia, extracts from classic Chinese texts, an overview of Chinese drug therapy, and notes on the Pen-Tsao Kang Mu or "Great Herbal," the "first" pharmacopoeia.

The Journal of Chinese Medicine
<http://www.jcm.co.uk/>

The Journal of Chinese Medicine is an English-language journal providing professional information on the entire field of Chinese medicine, including Chinese herbal medicine. Though the full text of the journal is not yet online, this site does provide tables of contents for the current issue and some back issues. Of particular note though are the sample full-text articles with useful information on Chinese herbal formulas used in the treatment of a variety of conditions. In addition, the online bookstore is a valuable guide to current texts on Chinese medicine.

Medicinal Plants of Native America Database (MPNADB)
<http://genome.cornell.edu/Botany/aboutmpnadb.html>

This database is based on a two-volume book of the same name published in 1986 by the Museum of Anthropology of the University of Michigan.[11] MPNAB contains "17,634 items representing the medicinal uses of 2,147 species from 760 genera and 142 families by 123 different Native American groups." Data can be retrieved using the botanical name, common name, and action/uses (e.g., cold remedy, burn dressing, heart medicine).

A Mini-Course in MEDICAL BOTANY
<http://www.ars-grin.gov/duke/syllabus/>

This resource provides useful lists of herbs used in China, Africa, Hawaii, India, and Arab and European countries (for more information about this site, see the corresponding entry in Chapter 4).

A Modern Herbal **(Mrs. Grieve)**
<http://www.botanical.com/botanical/mgmh/comindx.html>

This is a searchable hypertext version of the English herbal classic *A Modern Herbal: The Medicinal, Culinary, Cosmetic and Economic Properties, Cultivation and Folk-Lore of Herbs, Grasses, Fungi, Shrubs & Trees with their Modern Scientific Uses,* by Mrs. Maud Grieve. First published in 1931, this work is a little dated and thus does not reflect the most up-to-date studies. However, it is widely used as a reference book by many herbalists and provides extensive information for many herbs, including botanical information, medical properties, folklore, economic uses, dosage information, chemical composition, and harvesting techniques. It also includes information from British and U.S. pharmacopoeias, as well as some uses from nineteenth-century medicine. The index and text are searchable, but by common name only, and there are indexes of recipes and poisons.

The National Institute of Ayurvedic Medicine (NIAM)
<http://www.niam.com/corp-web/>

This is a comprehensive resource for basic information about the practice and beliefs of Ayurvedic medicine. The NIAM, located in Brewster, New York, is the largest resource for information on Ayurvedic medicine in the United States. Its research library has one of the largest collections of Ayurvedic literature in the country, with writings and research reports in English, Hindi, Sanskrit, Malayalam, Tamil, and several other dialects. Go to the medicinal plant page for profiles of plants from the Ayurvedic materia medica.

Native American Ethnobotany Database: Foods, Drugs, Dyes, and Fibers of Native North American Peoples
<http://www.umd.umich.edu/cgi-bin/herb>

This database is based on material compiled by Dr. Dan Moerman, Professor of Anthropology at the University of Michigan, Dearborn, and contains information on plants used by the Native North American peoples for foods, drugs, dyes, and fibers. See Figure 7.1 for a sample entry for *Echinacea angustifolia*.

Plantas Medicinais
<http://www.ciagri.usp.br/planmedi/temp.html>

This is a database of medicinal plants from the University of São Paulo in Brazil. Information is in Portuguese.

The Register of Chinese Herbal Medicine (United Kingdom)
<http://www.rchm.co.uk/>

The Register of Chinese Herbal Medicine was set up in 1987 to regulate the practice of Chinese herbal medicine (CHM) in the United Kingdom and to work with other organizations on issues relating to the practice and teaching of complementary and alternative medicine (CAM) therapies. This site has a collection of useful online articles, including such topics as the dispensing and prescribing of Chinese herbal medicines. It also provides information on the use of Western herbs in Chinese herbal medicine, a selection of formulas for herbal preparations, and abstracts of research articles from China.

FIGURE 7.1. Sample Record from the Native American Ethnobotany Database

Echinacea angustifolia DC.
; Asteraceae
Cheyenne Drug (Oral Aid)
Infusion of powdered leaves and roots taken or root chewed for sore mouth, gums or throat.
Grinnell, George Bird 1972 The Cheyenne Indians - Their History and Ways of Life Vol.2. Lincoln. University of Nebraska Press (188)

Tibetan Medicine Bibliography
  <http://www.uni-ulm.de/~jaschoff/bibli2.htm>

This is an annotated bibliography of resources in Tibetan medicine (1789 to 1995), corresponding to the print *Annotated Bibliography of Tibetan Medicine (1789-1995)* by Dr. Jürgen C. Aschoff, Department of Neurology, University of Ulm, Baden-Würtemberg, Germany.[12] It contains over 1,700 entries, listed alphabetically by author.

Tibetan Plateau Project (TPP)
  <http://www.earthisland.org/tpp/>

The Tibetan Plateau Project is dedicated to the protection of wildlife and plant resources in the Tibetan Plateau region, an area including portions of Bhutan, China, India, Nepal, Pakistan, and Tibet. TPP projects focus on endangered wildlife issues, medicinal plant resources, and public outreach activities. According to the TPP, the Tibetan Plateau region supports approximately 10,000 species of plants, with an estimated 1,000 having medicinal properties. TPP publishes a newsletter covering medicinal plant conservation and the practice of Tibetan medicine, as well as an electronic mailing list (see entry in Chapter 16).

The Tropical Plant Database
  <http://rain-tree.com/plants.htm>

Produced by Raintree Nutrition, Inc., this database lists plants found in the Amazon rainforests. Many of these, such as the Peruvian plant una de gato, or cat's claw *(Uncaria tomentosa),* are of increasing medical interest. Each plant file contains basic information such as taxonomic data, chemical constituents, ethnobotanical data, information on its uses in traditional medicine, images, and clinical research data with links to MEDLINE and other databases.

Tsumura & Co.: Leadership in Kampo Medicines (Information Center)
  <http://www.tsumura.co.jp/english/>

This is the Web site for Tsumura & Co., the world's largest producer of Kampo herbal products. It contains comprehensive information on Kampo, Japan's traditional herbal medicine system. Kampo is a holistic approach to health based on traditional Chinese medicine (TCM) and adapted to Japanese culture and experience. Select Kampo Today to view information on the background and principles of Kampo and Kampo therapeutics, including laboratory and clinical research findings, scholarly meetings, and industry and regulatory trends.

Unani Herbal Healing
<http://www.unani.com/>

Unani is a traditional healing system derived from Greco-Arabic medicine; it is primarily a type of herbal medicine that is currently practiced in India and Pakistan. This Web site, provided by the American Institute of Unani Medicine, contains comprehensive information on the theory and practice of Unani medicine. At the home page, select Unani Herbology for an overview of this topic. For specific information on herbs, including the twenty most important herbs used in Unani and the formulation of herbal preparations, go to Quick Links.

The World Health Organization (WHO)
<http://www.who.int/>

WHO Collaborating Centers

The World Health Organization (WHO) came into being in 1948 as the health agency of the United Nations. One of its core objectives is the establishment of a high standard of health for all peoples, in all countries. Realizing the importance of indigenous medicine traditions in many developing countries, WHO strongly advocates cooperation between traditional and modern health practitioners. Through its Traditional Medicine Programme, WHO is gathering information on medical herbs that are widely used in primary care settings throughout the world, while also promoting programs to establish research guidelines for evaluating their safety and efficacy and to educate the public and health care practitioners.

WHO has established nineteen Collaborating Centers in ten countries: Belgium, China, Democratic People's Republic of Korea, Italy, Japan, Republic of Korea, Romania, Sudan, Vietnam, and the United States. The two centers in the United States are the College of Pharmacy,

University of Illinois at Chicago, which produces the NAPRALERT database (see the corresponding entry in Chapter 13), and the Institute for Advanced Research in Asian Science and Medicine (IARASM) in Brooklyn, New York, publisher of the *American Journal of Chinese Medicine* (AJCM).

WHO Publications: Pharmaceuticals, Biologicals
<http://www.who.int/dsa/cat98/phar8.htm>

WHO Publications: Traditional Medicine
<http://www.who.int/dsa/cat98/trad8.htm>

The WHO publishes a number of useful fact sheets and monographs on herbal medicine, such as "Guidelines for the Appropriate Use of Herbal Medicines," "Medicinal Plants in China," "Medicinal Plants in Viet Nam," and "Quality Control Methods for Medicinal Plant Materials." Several are available online as full text. To see a list of relevant WHO publications, with descriptions, use the URLs listed here.

Publication has just begun of a series of twenty-eight monographs covering the quality control, traditional uses, and clinical uses of selected medicinal plants.[13]

WHOLink: WHO Library Home Page
<http://www.who.int/hlt>

The WHO site home page also provides a link to the WHO library catalog. This catalog includes not only citations for monographs in its collection, but also provides access to WHO technical documents, some of which are available electronically as full-text documents. To search the library catalog, click on WHOLIS (Library Catalogue).

Chapter 8

Laws, Regulations, and Standards

The nice thing about standards is that there are so many of them to choose from.

Andrew S. Tanenbaum

DRUG LAWS AND REGULATIONS

The Food and Drug Administration (FDA) is the U.S. government agency responsible for ensuring the health and safety of the American public by enforcing laws and regulations related to pharmaceutical drugs. Before they reach the marketplace, potential drugs must go through several phases of testing before they are approved. The U.S. system is perhaps the most rigorous in the world. On average, it costs a pharmaceutical company close to $400 million to get one new drug from the laboratory to the pharmacy shelf, and the whole process takes an average twelve years to complete.[1] This involves a complicated system of clinical trials designed to assess the drug's safety and efficacy. Unfortunately, little information is available to accurately assess the safety and effectiveness of many herbal therapies. A pharmaceutical manufacturer has little or no economic incentive to invest in expensive basic research programs and clinical trials when many naturally occurring plant extracts cannot be patented. Only recently have some human clinical trials been undertaken, usually funded by government agencies, and then for just a limited number of widely used herbs, such as echinacea, St. John's wort, and saw palmetto.

THE DIETARY SUPPLEMENT HEALTH
AND EDUCATION ACT (DSHEA) OF 1994

The current U.S. approach to the regulation of herbal products is rather complex and generally considered to be unsatisfactory. Until 1994, the FDA classified most herbal remedies as either "food additives" or "drugs," and manufacturers had to meet strict FDA standards before placing their products on the U.S. market. This changed in 1994 when Congress enacted the Dietary Supplement Health and Education Act (DSHEA), which placed herbs together with vitamins and minerals in a category know as "dietary supplements." DSHEA defines a dietary supplement as follows:

> a product (other than tobacco) intended to supplement the diet that bears or contains one or more of the following dietary ingredients: a vitamin, a mineral, an herb or other botanical, an amino acid, a dietary substance for use by man to supplement the diet by increasing the total daily intake, or a concentrate, metabolite, constituent, extract or combination of any (of these ingredients).

DHSEA effectively means that manufacturers can now market herbal remedies and other supplements without first obtaining FDA approval of their safety and efficacy. Under DSHEA, manufacturers do not have to demonstrate that their products are effective, or even that they are safe, and they are not required to report side effects. The FDA must also prove that a supplement is harmful before it can be banned.

DHSEA also addresses labeling requirements, specifying that although manufacturers can make claims about how a product affects the body's "structure and function," they cannot claim that it prevents, treats, or cures a disease or medical condition. Manufacturers cannot, therefore, make a therapeutic claim that cranberry juice "can prevent urinary tract infections," but they can say that it "helps maintains a healthy urinary tract."

CONSISTENCY AND STANDARDIZATION

In addition to the controversies over their regulatory status, some concern exists about the identity, purity, quality, and strength of many herbal preparations currently on the market. Since the FDA does not regulate quality control measures for herbal products, the consumer must rely on the integrity of the manufacturer. Unfortunately, numerous instances of misbranding, adulteration, and contamination of herbal products, sometimes resulting in poisoning, have been reported.[2,3] Consistency and standardization of extracts are other areas that need to be examined. A 1995 study by *Consumer Reports* of ten popular brands of ginseng showed that the amount of active ingredient varied considerably, with some actually having no ginseng in them.[4] Another study of feverfew *(Tanacetum parthenium)* preparations used to treat migraine found that many commercial brands had no detectable levels of parthenolide, the active component in feverfew.[5] To complicate matters, some herbal preparations may have several active constituents—so which one should be standardized? Sometimes the active ingredients have not even been identified. However, without adequate guidelines and regulations, consumers cannot be certain that an herbal product contains sufficient quantities of an active ingredient for it to have any therapeutic effect.

Most manufacturers of herbal preparations have their own set of criteria to ensure the quality of their products, and these criteria differ from company to company. A section of the herbal industry is, however, trying to address this problem by developing standard testing methods. One such initiative is the Methods Validation Program (MVP), "an international effort to validate and make available analytical methods that will meet the demand for global consistency in the testing of botanicals" (see the entry in this chapter for the Institute for Nutraceutical Advancement [INA]). The FDA is also in the process of proposing new good manufacturing practice (GMP) regulations for herbal preparations, governing their preparation, packaging, and storage under conditions that ensure their safety. Equivalent legislation is being considered by other countries and by an international body known as the Codex Alimentarius Commission (see the corresponding entry in this chapter).

EUROPE AND GERMANY'S COMMISSION E MONOGRAPHS

Whereas widespread consumer interest in medicinal herbs is a relatively new phenomenon in the United States, herbs have been widely used in several European countries for many years. They play a particularly important role in German medicine, in which they are not only used for self-medication but also widely prescribed by medical doctors. Since 1993, medical students must successfully complete the phytotherapy portion of their board examinations as a precondition for practicing medicine! Due to the significant number of entries for herbal teas and similar preparations, the *Rote Liste* (Red List), the German equivalent of the *Physicians' Desk Reference* (PDR), reads like a cross between the PDR and a health food catalog.

In Germany and other European countries the term *phytomedicine* has been in use for many decades ("phyto" originates from the Greek *phyton,* meaning plant). It is a recognized category of plant-derived drug products, referring to therapeutic products prescribed (or recommended) by physicians. In recognition of the significant role that phytomedicines play in that country's health care system, the German government has developed a unique mechanism to provide health practitioners and consumers with accurate information on their safety and efficacy. The German Medicines Act (Arzneimittelgesetz) of 1976 provided for the issuance of standard licenses for herbal products, subject to the publication of a monograph providing qualitative and quantitative information. The German Federal Health Agency (Bundesgesundheitsamt [BGA]), equivalent to the U.S. FDA, established a Commission E to evaluate herbal preparations (hence the "Commission E" of the work's title) and to prepare monographs. The commission was given the task of examining data on around 1,400 different herbs, including results from clinical trials, information from the scientific literature, and the collective conclusions of the major German medical associations. Results were published as Commission E monographs in the German *Bundezanzeiger* (Federal Gazette). About 300 monographs have appeared so far, covering most of the economically important herbal remedies in Germany. The information was designed to be printed inside product packaging sold in German pharmacies, so each monograph provides the description of the herb, formulation, pharmacological properties, clinical data, recommended doses, approved uses, contraindications, precautions, and side effects. Of perhaps equal significance is the information included on herbs that were found not to be effective.

In response to the urgent need in the United States for more authoritative information, the American Botanical Council (ABC) sponsored an English translation of the Commission E monographs. These were published in the summer of 1998 as *The Complete German Commission E Monographs: Therapeutic Guide to Herbal Medicines.*[6] Both traditional herbal practitioners and mainstream physicians have welcomed their publication in the United States and regard this as a major advance in herbal medicine. In addition to the importance of the monographs themselves, many are looking to the German system as a possible model for the future regulation of herbal products in the United States.

Government agencies in North America and in several European countries are currently trying to develop a regulatory framework that would satisfy health professionals, consumers, and the herbal industry. Recent developments in North America include the establishment of the Office of Dietary Supplements (ODS) in the United States and the Office of Natural Health Products (ONHP) in Canada.

WEB SITES OF GOVERNMENT AND PRIVATE REGULATORY AGENCIES

Botanicals Generally Recognized As Safe
  <http://www.ars-grin.gov/duke/syllabus/gras.htm>

Summary of all GRAS Notices
  <http://vm.cfsan.fda.gov/~dms/opa-gras.html>

In 1938 Congress passed the federal Food, Drug and Cosmetic Act in response to an incident involving the marketing of "Elixir of Sulfanilamide," a medicinal syrup containing the toxic compound diethylene glycol (a component of today's automobile antifreeze). Nearly 100 people were killed before the product was withdrawn from the market.[7] The 1938 act was a milestone in drug regulation in the United States. For the first time, manufacturers were required to provide drug labels listing directions for use and appropriate warnings. It also ushered in the modern era of drug regulation by stipulating that new drugs be tested for safety before being marketed. Future amendments to the 1938 law ensured that no substance could be introduced into the food supply without a prior determination of its safety. The 1958 Food Additives Amendment required a manufacturer of new food additives to establish safety, with the exception of a category of food additives that over time have been generally recognized as safe (GRAS) for human consumption. Further information on what constitutes GRAS status can be found in the Code of Federal Regulations (CFR), Title 21, section 170.30: "Eligibility for classification as generally recognized as safe (GRAS)" (see the CFR entry in this chapter). Updates to the GRAS list can be found at the Inventory of GRAS Notices. See Table 8.1 for a summary of U.S. drug legislation affecting herbal remedies.

About 250 herbs are included on the GRAS list by the FDA because of their use in liqueurs and as components of natural flavorings. This online list is a compilation of all GRAS botanicals accepted by the FDA for use as food additives and is part of James Duke's A Mini-Course in MEDICAL BOTANY (see the corresponding entry in Chapter 4).

Code of Federal Regulations (CFR)
<http://www.access.gpo.gov/nara/cfr/>

Federal Register
<http://www.gpo.gov/su_docs/aces/aces140.html>

The Code of Federal Regulations (CFR) contains the text of public regulations issued by agencies of the federal government. Proposed

TABLE 8.1. Major U.S. Legislation Affecting Herbal Products

1906	The Pure Food and Drugs Act prohibits interstate commerce in misbranded and adulterated foods, drinks, and drugs.
1938	The Federal Food, Drug, and Cosmetic (FDC) Act requires all new drugs to be proven safe before marketing, starting a new system of drug regulation.
1958	The Food Additives Amendment requires manufacturers of new food additives to establish safety and prohibits the approval of any food additive shown to induce cancer in humans or animals. The FDA publishes in the *Federal Register* the first list of Substances Generally Recognized As Safe (GRAS). The list contains nearly 200 substances.
1994	The Dietary Supplement Health and Education Act (DSHEA) establishes specific labeling requirements and provides a basic regulatory framework for dietary supplements. This act defines "dietary supplements" to include herbs.
1997	The Food and Drug Administration Modernization Act mandates wide-ranging reforms in FDA practices. It includes measures to regulate health claims for foods—including dietary supplements.

regulations, issued so recently that they are not yet in the CFR database, can be found in the *Federal Register*. CFR is divided into fifty titles. The FDA's issued regulations on the implementation of DSHEA are codified in Title 21, "Food and Drugs of the Code of Federal Regulations (CFR)." Together with the statutes in DSHEA, these specific regulations govern the dietary supplement area in the United States and cover such items as labeling and package inserts.

Codex Alimentarius Commission
<http://www.fao.org/WAICENT/FAOINFO/ECONOMIC/ ESN/codex/>

Codex alimentarius is Latin for "nutrition code." The Codex Alimentarius Commission is a subsidiary body of the United Nations Food and Agriculture Organization (FAO) and the World Health Organization (WHO). Established in 1963, it seeks to develop an international code of food quality standards to protect the health of consumers and to ensure fair practices in food trade. The Codex Alimentarius has become the seminal global reference point for consumers, food producers and processors, national food control agencies, and the international food trade. These are all voluntary standards, however, and there is no direct obligation on member governments to apply Codex standards.

The Codex Committee on Nutrition and Foods for Special Dietary Use (CCNFSDU) is charged with examining dietary supplements, including herbal medicines. The Codex committee decided to develop a guidance document addressing the sale of potentially harmful herbs and botanical preparations. This, however, was opposed by the U.S. delegation, so no action has as yet been taken.

ConsumerLab.com
<http://www.consumerlab.com>

ConsumerLab.com is an independent testing company, in White Plains, New York, that provides consumer information and independent evaluations of health products. It is currently testing popular herbal products to see if they meet scientific standards for potency, purity, and consistency. (See the corresponding entry in Chapter 11 for more information.)

Enhancing the Accountability of Alternative Medicine (Report)
<http://www.milbank.org/mraltmed.html>

The Milbank Memorial Fund is an endowed national foundation that supports nonpartisan analysis and study of, and research on, significant issues in health policy. This report, published in January 1998, deals with the use of complementary and alternative medicine (CAM) in the United States today, with a focus on how to increase the accountability of CAM practitioners. It covers the major types of CAM practice with information on regulation and reimbursement.

The European Phytojournal
<http://www.exeter.ac.uk./phytonet/phytojournal/>

This is the official journal of the European Scientific Cooperative on Phytotherapy (ESCOP) (see the following ESCOP entry for details). Registration is required, though the journal is free. Items include papers from ESCOP international symposia, regulatory reviews, updates of research and clinical literature, and safety reviews.

European Scientific Cooperative on Phytotherapy (ESCOP)
<http://info.ex.ac.uk/phytonet/escop.html>

The European Scientific Cooperative on Phytotherapy (ESCOP) is an umbrella organization of national associations for phytotherapy established in 1989 to advance scientific studies of herbal preparations and to assist with the harmonization of their regulatory status in European countries. The cooperative's main objective is to develop therapeutic herbal monographs. The first two collections, each containing ten monographs, were published in March 1996. ESCOP members now include associations from the majority of countries within the European Union and some from a number of non-EU countries, including the American Botanical Council. (See the corresponding entry in Chapter 5; see also the previous related entry for *The European Phytojournal*.)

Food and Drug Administration (FDA)
<http://www.fda.gov/>

FDA Center for Drug Evaluation and Research (CDER)
<http://www.fda.gov/cder/>

FDA Center for Food Safety & Applied Nutrition (CFSAN)
<http://vm.cfsan.fda.gov/list.html>

FDA/CFSAN Dietary Supplements
<http://vm.cfsan.fda.gov/~dms/supplmnt.html>

The FDA is the federal organization designed to evaluate and regulate a wide variety of products for human and animal use. Its mission is "to promote and protect the public health by helping safe and effective products reach the market in a timely way, and monitoring products for continued safety after they are in use." FDA-regulated products range from medical devices, such as hearing aids, to prescription drugs and over-the-counter (OTC) drug products. FDA experts weigh evidence assessing the effectiveness and the benefits of each product against any potential risks and side effects; they then decide whether to approve the product for marketing.

The FDA is divided into several branches, each with its own responsibilities. One division, the Center for Drug Evaluation and Research (CDER), is responsible for developing and enforcing safety standards and regulating the drug approval process. From the FDA's home page, select Food (to access the Center for Food Safety & Applied Nutrition); then from Program Areas choose Dietary Supplements to locate details on DSHEA and to view FDA information on reported adverse reactions and fraudulent claims.

As might be expected, the FDA Web site hosts many documents of relevance to the regulation of herbal drugs. Navigate using the Site Map or simply use the search engine by entering appropriate keywords: note that relevant FDA documents use the term *botanical.* The FDA's Dietary Supplements site is a large collection of selected resources specifically designed to answer questions about the regulation of herbs and other dietary supplements. It includes extensive information on DSHEA, its implementation, ramifications, and up-to-date items concerning regulatory activities and proposed actions.

Guidance Documents—Botanical Drug Products (FDA Guidance for Industry)
<http://www.fda.gov/cder/guidance/>

Guidance documents represent the FDA's current views on a particular topic. This document provides guidance on submitting investigational new drug applications (INDs) for herbal drug products in the United States. To locate this report, go to the Chemistry (Draft) section.

Institute for Nutraceutical Advancement (INA)
<http://www.nutraceuticalinstitute.com/>

Physicians and pharmacists are now requesting that the herbal industry develop quality and consistency standards for herbal products. The INA is a division of Denver-based Industrial Laboratories, an independent laboratory providing analytical and consulting services to the natural products industry. Its work on testing methods for plant materials is sponsored by a consortium of manufacturers and suppliers of herbal products dedicated to the development of standard methods for analyzing dietary supplements. Notable participating organizations include the American Botanical Council (ABC), the American Herbal Pharmacopoeia (AHP), the Association of Official Analytical Chemists (AOAC), the Council for Responsible Nutrition (CRN), the Herb Research Foundation (HRF), and the FDA.

The INA has launched a Methods Validation Program (MVP), described as "an international . . . project designed to select, validate and publish scientific methods for use in analyzing raw botanical materials." To date, the INA has published analytical methods for over fifteen herbal compounds, including ginkgoterpenoids in *Ginkgo biloba,* ginsenosides in *Panax ginseng* and American ginseng, polyphenols in echinacea, hyperforin in St. John's wort, gingerols and shogaols in ginger *(Zingiber offinale),* and silymarins in milk thistle. These are available online.

Office of Dietary Supplements (ODS)
<http://dietary-supplements.info.nih.gov/>

Section 13 of the Dietary Supplement Health and Education Act (DSHEA) of 1994 provided for the creation of an Office of Dietary Supplements (ODS) within the National Institutes of Health. The ODS supports research into dietary supplements and cosponsors clinical studies of herbal medicines. Part of its mission is to promote the dissemination of research results to health professionals and the public, developing two important databases: Computer Access to Research on Dietary Supplements (CARDS), a database of existing and ongoing dietary supplements research currently supported by federal agencies, and the International Bibliographic Information on Dietary

Supplements (IBIDS), a database of published, peer-reviewed international scientific literature on dietary supplements (see the IBIDS entry in Chapter 13).

The ODS Web site provides access to the IBIDS database, as well as information on the office's funding opportunities, research activities, meetings, conferences, and other events. In 2001 it published the first edition of an *Annual Bibliography of Significant Advances in Dietary Supplement Research,* which can be downloaded from the Web site. Interested individuals can also subscribe to an ODS mailing list that provides updates on ODS activities and additions to the Web site.

Office of Natural Health Products (ONHP) (Canada)
<http://www.hc-sc.gc.ca/hpb/onhp/>

Over 50 percent of Canadians now use some type of "natural" health product, such as herbs, vitamins, and mineral supplements; traditional Chinese or Ayurvedic medicine; and homeopathic preparations.[8] In 1999, in response to concerns from consumers, health care providers, and industry, the Canadian government established the ONHP to develop a new regulatory framework for these products. The new office is designed to "provide Canadian consumers with the assurance of safety while enhancing consumer access and choice to a full range of natural health products."

Overview of Legislative Development Concerning Alternative Health Care in the United States
<http://www.healthy.net/public/legal-lg/regulations/fetzer.htm>

This is a comprehensive look at regulation of CAM practices in the United States and specific federal and state legislation affecting this. The author is David M. Sale, JD, LLM, and the report was produced as a research project for the Fetzer Institute. Select the Herbology and Regulation of Dietary Supplements links for the most relevant sections.

The Fetzer Institute, in Kalamazoo, Michigan, is named after John Earl Fetzer (1901-1991). It is a nonprofit private foundation supporting programs that explore connections between the "physical, psychological, social, and spiritual dimensions of life, and how understandings in these areas can improve health, foster growth, and better the human condition."

Thomas: Legislative Information on the Internet
<http://thomas.loc.gov/>

Laws and regulations involving herbs and other dietary supplements are in a state of flux at the moment. DHSEA is generally considered to be very inadequate and the FDA is being pressured to change its approach. This is likely to be a dynamic legislative area over the next decade. Thomas is provided by the Library of Congress and was set up at the request of the U.S. Congress to make legislative information freely available on the Internet. This site provides access to the full text of legislation (bills and laws) from the 101st Congress (1989-1990) to the present and is useful for tracking legislation involving herbs and other dietary supplements.

The United States Pharmacopeial Convention/
United States Pharmacopeia (USP)
<http://www.usp.org/>

Founded in 1820, the United States Pharmacopeial Convention, Inc., is a volunteer, not-for-profit organization composed of health professionals, representatives from the pharmaceutical industry, government agencies, and consumer organizations. This organization sets the standards for strength, quality, purity, packaging, and labeling for medical products in the United States. These standards are enforceable by the FDA, are recognized by more than thirty-five other countries, and are the official reference guide for manufacturers and analytical laboratories. More than 3,700 standards monographs are published in the *United States Pharmacopeia* and *National Formulary* (USP/NF), the official drug standards compendia.[9]

In an effort to promote industrial standards on purity, strength, quality, and labeling, the USP has formed a Subcommittee on Natural Products to create monographs on herbal products for publication in the USP and NF. Manufacturers who meet these national standards will be able to put "NF" or "USP" on the label. So far monographs have appeared for valerian, St. John's wort, saw palmetto, milk thistle, Asian ginseng, ginkgo, ginger, cranberry, chamomile, and garlic. At the home page, select Dietary Supplements for information on the status of other herbal monographs currently in preparation.

U.S. Department of Commerce (DOC)
<http://www.doc.gov/>

The DOC is the federal government department charged with promoting U.S. economic development and technological advancement. The DOC impacts the herbal industry in the United States through its involvement in international trade, import/export control, and promoting access to the international pharmaceutical and nutritional supplements markets. It produces a variety of reports on the economic and business aspects of herbal commerce. Many of these are evaluations of the herbal product markets in a wide range of foreign countries, with valuable information on those countries' regulatory frameworks.

World Health Organization (WHO)
<http://www.who.org>

WHO Publications: Traditional Medicine
<http://www.who.int/clsa/cat98/trad8.htm>

The World Health Organization (WHO) is actively involved in promoting programs to establish research guidelines for procedures for assessing the identity, purity, and content of medicinal plant materials. One of its publications, "Guidelines for the Appropriate Use of Herbal Medicines," is a report of the findings and recommendations of a working group convened to prepare guidelines for the use of herbal medicines in Western Pacific countries, with information on legal and other options available for the regulation of practitioners, manufacturers, and the distribution system. Another of its publications, *Quality Control Methods for Medicinal Plant Materials,* is a collection of recommended test procedures for assessing the identity, purity, and content of medicinal plant materials.[10,11] Go to the WHO Publications page for Traditional Medicine to find information for relevant WHO reports. (For more information about the WHO, see the corresponding entries in Chapter 7.)

Chapter 9

Clinical Evidence and Clinical Trials

It is wrong always, everywhere and for everyone to believe anything upon insufficient evidence.

W. K. Clifford
British philosopher, circa 1876

DO HERBS WORK?

In the United States, drugs are submitted to rigorous scientific testing to determine their efficacy and safety. Overseen by the Food and Drug Administration (FDA), this involves a series of clinical trials with reports published in quality peer-reviewed medical journals. Many health professionals believe that randomized controlled trials (RCT) are the only way to evaluate the effectiveness of pharmaceutical products. In an RTC, participants are assigned by chance to receive either an experimental or a control treatment and any differences seen in the groups at the end can be attributed to the difference in treatment alone, and not to bias or chance.

Mainstream Western medicine regards the RCT as the "gold standard," that is, the standard against which other procedures are measured for providing scientific evidence that a drug actually works. However, it is not the only way evidence can be obtained. Case studies, experiential reports, and observational findings of practitioners may provide valuable insights. Long-standing traditional or indigenous systems of medicine, for example, Chinese, Ayurvedic, and Native American, are empirical systems of knowledge, handed down from generation to generation. Traditional knowledge of herbs is often based upon empirical evidence or "trial and error." Such knowledge has practical validity, even though most traditional practices have not been studied in randomized controlled clinical trials. Many health practitioners believe that the RCT is not always the most appropriate research methodology for alternative and complementary treatments.

EVIDENCE-BASED MEDICINE (EBM)

The randomized controlled clinical trial is the bedrock of what is now called evidence-based medicine (EBM). EBM is "the conscientious, explicit, and judicious use of current best evidence in making decisions about the care of the individual patients."[1] Many medical schools are now reorganizing their curricula to teach students how to take an evidence-based medicine approach to solving medical problems. This involves finding, appraising, and using the most up-to-date research findings as the basis for clinical decisions, usually based on information gathered from clinical trials. Evidence-based practice aims to move beyond anecdotal clinical experience by bridging the gap between research and the practice of medicine.

The current trend toward EMB has ramifications for complementary and alternative medicine (CAM) remedies and techniques. Critics of CAM insist that an evidence-based medicine approach be used to establish that these therapies actually work. Federal agencies such as the National Center for Complementary and Alternative Medicine (NCCAM) and the Office of Dietary Supplements (ODS) are thus spending large sums of money to conduct clinical trials for herbal therapies, and corporate sponsorship of research is on the rise. The Cochrane Collaboration, serving as a major resource for EMB, has even established a special complementary medicine group to locate and analyze clinical trail data for CAM therapies (see entry in this chapter).

Since a great deal of herbal knowledge is based on traditional knowledge and empirical or "anecdotal" evidence gathered over many years, herbal practitioners criticize allopathic medicine for insisting that a rigorous clinical trial is the only valid method for evaluating safety and efficacy. They complain that conventional medicine itself continues to use many treatments that have not been tested in clinical trials and that physicians have acted largely on the basis of clinical intuition, anecdotal evidence, personal experience, or help from more experienced colleagues.

Exactly how much of contemporary medicine is based on science is still up for debate. A 1983 report from the U.S. Office of Technology Assessment (OTA) suggested that "only 10% to 20% of all procedures currently used in medical practice have been shown to be efficacious by controlled trial," [2] and according to physician and mathematician David M. Eddy, MD, PhD, "as few as 15% of medical decisions are based on the results of rigorous evidence."[3]

WEB SITES PROVIDING INFORMATION ON CLINICAL EVIDENCE AND TRIALS

Alternative Medicine Review:
A Journal of Clinical Therapeutics
<http://www.thorne.com/altmedrev/>

This peer-reviewed journal provides clinically relevant articles, abstracts, and literature reviews on natural therapeutics aimed at the CAM practitioner. A fee-based subscription is required to view the complete text, but titles and abstracts can be browsed online for free.

American Medical Association (AMA)
<http://www.ama-assn.org/>

The AMA is the leading medical society in the United States, developing and promoting standards in medical practice, education, and research. Its flagship journal is the *Journal of the American Medical Association* (JAMA). In recent years it has paid close attention to CAM therapies, particularly herbal medicine, and articles published in JAMA have been important in bringing this topic to the attention of American physicians.

Association of Natural Medicine Pharmacists (ANMP)
<http://www.anmp.org/>

The ANMP's mission is "to provide pharmacist education and certification on natural medicines." Its founder and president, Constance E. Grauds, RPh, is an Assistant Professor of Clinical Pharmacy at the University of California, San Francisco; Natural Medicine Editor for *Pharmacy Times;* and a reviewer for Facts and Comparison's *Review of Natural Products*.[4] The AMMP offers pharmacists several education programs, including a certificate program in Phytomedicinals and Alternative Medicine. Documents available from the site include herbal monographs and articles from its newsletter, *The Source.*

bmj.com Collected Resources: Complementary Medicine
<http://www.bmj.com/cgi/collection/complementary_medicine>

This is a collection of CAM articles from the *British Medical Journal* (BMJ). Access to the entire BMJ Web site is free to all, and this collection includes journal articles, editorials, and letters from the journal. In 1999, the BMJ published of articles on CAM therapies in their Clinical Review series, including one on herbal medicine.

ClinicalTrials.gov
<http://clinicaltrials.gov/>

Developed by the National Institutes of Health (NIH), through the National Library of Medicine (NLM), ClinicalTrials.gov provides patients, families, and members of the public easy access to information about the location of clinical trials, their design and purpose, criteria for participation, and, in many cases, further information about the disease and treatment under study. The site also features links to individuals responsible for recruiting participants for each study. Use the Experimental Treatment box on the Focused Search link from the home page to search for herbs by common or botanical name. See Figure 9.1 for sample search results for echinacea.

The Cochrane Collaboration
<http://www.cochrane.org/>

Abstracts of Systematic Reviews
<http://www.cochrane.org/cochrane/revabstr/mainindex.htm>

Database of Abstracts of Reviews of Effectiveness (DARE)
<http://nhscrd.york.ac.uk/darehp.htm>

The Cochrane Collaboration is an international network committed to preparing, maintaining, and disseminating systematic reviews

FIGURE 9.1. Sample Record from ClinicalTrials.gov Database

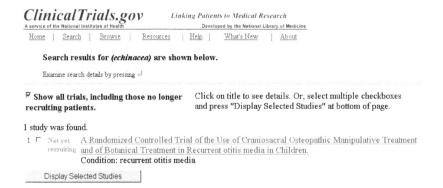

of research on the effects of health care. It is considered by many health professionals to be the primary resource for locating evidence-based information. Its research is published in four databases: the Cochrane Database of Systematic Reviews (CDSR); the Database of Abstracts of Reviews of Effectiveness (DARE); the Cochrane Controlled Trials Register (CENTRAL/CCTR); and the Cochrane Review Methodology Database. From the Site Index, select the Cochrane Library link for information on accessing sections that require a paid subscription.

The Cochrane Database of Systematic Reviews (CDSR) is a collection created and maintained by the Cochrane Collaboration. A systematic review aims to be a comprehensive overview of all the literature on a particular subject. Each review attempts to synthesize the results and conclusions of any primary studies in the field that satisfy certain standards. A paid subscription through The Cochrane Library is required to view the reviews themselves, but abstracts are available online without charge and can be browsed or searched. The work of writing and updating the systematic reviews is carried out by around fifty Collaborative Review Groups, supported and guided by other subgroups. For complementary and alternative medicine (CAM), there is a Complementary Medicine Field Group that searches the literature and other sources for relevant CAM studies and provides other information to the main review groups. Reviews have been published for St. John's wort *(Hypericum perforatum),* in the treatment of depression, and for saw palmetto *(Serenoa repens),* in the treatment of benign prostatic hyperplasia (BPH).

The Database of Abstracts of Reviews of Effectiveness (DARE) includes structured abstracts of systematic reviews from around the world that have been critically appraised by reviewers at the National Health Service (NHS) Centre for Reviews and Dissemination at the University of York, England. DARE search results will include some Cochrane reviews plus others from the medical literature.

Enhancing the Accountability of Alternative Medicine (Report)
<http://www.milbank.org/mraltmed.html>

This report deals with the use of CAM therapies in the United States today, with a focus on their cost, reimbursement issues, and the evaluation of CAM therapies (see Chapter 8 for more information).

HerbMed
<http://www.herbmed.org/>

HerbMed is a project of Alternative Medicine Foundation, Inc., a nonprofit group providing evidence-based information resources and reliable information about alternative medicine. The HerbMed database provides comprehensive profiles of the most widely used herbs, including summaries of human clinical studies, with hyperlinks to the citation, and abstracts in the PubMed database. Other information includes adverse interactions, traditional uses, chemical constituents, mechanism of action, and descriptions of commercial and home-based preparations of herbal medicines.

Medical Herbalism: A Journal for the Clinical Practitioner
<http://www.medherb.com/MHHOME.SHTML>

Medical Herbalism is a quarterly journal of clinical herbalism founded to "strengthen the herbal practitioner, to preserve and develop the science and art of herbal medicine, and to promote communication and sharing of clinical methods and experiences." Its editor, Paul Bergner, teaches at the Rocky Mountain Center for Botanical Studies, in Boulder, Colorado (see the corresponding entry in Chapter 6). Though a paid subscription is required for the journal itself, the site provides the full text of sample articles and tables of contents for back issues.

Office of Dietary Supplements Publications:
Annual Bibliography of Significant Advances in Dietary Supplement Research
<http://ods.od.nih.gov/publications/publications.html>

This is a publication of the Office of Dietary Supplements (ODS) (see the ODS entry in Chapter 8). Copies of the bibliography may be downloaded from the ODS Web site.

Pharmacy Times
<http://www.pharmacytimes.com/>

Pharmacy Times is a retail publication provided free of charge to qualified pharmacists and other pharmacy professionals. The Herbal Q & A link from the home page leads to a collection of herbal questions posted and answered by pharmacists.

PubMed Clinical Query
<http://www.ncbi.nlm.nih.gov:80/entrez/query/static/clinical. html>

This clinical queries interface to the MEDLINE database is one approach to finding evidence-based medicine journal articles. It is aimed at the clinician who is interested in retrieving only citations for clinical studies. The system uses built-in search filters for therapy, diagnosis, etiology, or prognosis. For information on PubMed, see the corresponding entry in Chapter 13.

Chapter 10

Adverse Effects, Adverse Reactions, and Drug Interactions

If the whole Materia Medica, as now used, could sink to the bottom of the sea, it would be all the better for mankind, and all the worse for the fishes.

Oliver Wendell Holmes

HERBAL REMEDIES ARE NOT RISK FREE

The increased use of herbal preparations is causing concern among doctors and pharmacists due to the widespread misconception among the American public that because herbal remedies are "natural," they are free from the adverse effects or reactions associated with conventional drugs. Adverse effects, adverse reactions, and drug interactions are all areas of concern. An adverse effect is "[a]ny unintended effect of a pharmaceutical product occurring at doses normally used in man which is related to the pharmacological properties of the drug," while an adverse reaction, is "[a] response to a drug which is noxious and unintended, and which occurs at doses normally used in man for the prophylaxis, diagnosis, or therapy of disease, or for the modification of physiological function."[1]

Many people continue to take herbal products unaware of some of the risks involved. Herbal products containing *Ephedra sinica,* or ma huang, have been associated with adverse cardiovascular problems, seizures, and even death.[2] Ginkgo biloba extract, used to improve cognitive functioning, has been reported to cause spontaneous bleeding.[2] Serious interactions have been reported between herbs and conventional drugs, and even the relatively benign St. John's wort has now been reported to demonstrate serious interactions with certain HIV protease-inhibitor drugs used in the treatment of HIV/AIDS patients.[3]

Under current U.S. law, before new drugs can be approved for marketing, a pharmaceutical company must provide evidence that they are not only effective but nontoxic and safe to use. Clinical trials regulated by the FDA determine safety and efficacy, at what dosage the drug works best, and its side effects. Unfortunately, no drug is completely free of side effects, so there is always some risk of an adverse reaction. Often the FDA must determine whether the benefits outweigh the risks. Serious adverse drug reactions occur more frequently than is commonly thought. A 1998 JAMA paper claimed that adverse drug reactions are a serious problem in U.S. hospitals, each year causing the deaths of up to 100,000 patients.[4]

The manufacturers of herbal products are currently not required to submit proof of safety and efficacy to the FDA before bringing a product to market. For this reason, the adverse effects and drug interactions associated with herbal remedies are largely unknown. Physicians and pharmacists are increasingly likely to encounter patients who are using herbal remedies. They need to be aware of the adverse effects of herbal remedies and the possibility of interactions with prescription drugs.

An increasing amount of information on herb-associated adverse reactions and herb-drug interactions is appearing in the journal literature. To help make this information more readily available to doctors and physicians, programs such as the FDA's Medwatch system and the European PhytoNET network have been established specifically to collect such information from health professionals.

WEB SITES THAT DISCUSS
ADVERSE HERB REACTIONS
AND DRUG-HERB INTERACTIONS

American Society of Anesthesiologists
<http://www.asahq.org/>

What You Should Know About Herbal Use and Anesthesia
<http://www.asahq.org/PublicEducation/insidherb.html>

Some popular herbal remedies can be dangerous if taken before surgery, either prolonging the sedative effect of anesthesia or increasing bleeding during surgery and causing fluctuations in blood pressure.[5] The Web site for the American Society of Anesthesiologists (ASA) provides the patient with information on herbal interactions and what to tell the physician. At the home page, select Public Education, then the Additional Information subcategory, then go to the document What You Should Know About Herbal Use and Anesthesia.

American Society of Health-System Pharmacists (ASHSP)
<http://www.ashp.org/>

The American Society of Health-System Pharmacists (ASHP) is a national professional association that represents pharmacists who practice in hospitals or in other sections of the health care sector, such as health maintenance organizations, long-term care facilities, and home care agencies. Important society publications include the *American Journal of Health-System Pharmacy.* From the home page, under Products and Services, select Continuing Education to find the clinical review "Unsafe and Potentially Safe Herbal Therapies" and other relevant documents concerning herbal preparations.

Institute for Safe Medication Practices (ISMP)
<http://www.ismp.org/>

The ISMP is a nonprofit organization working closely with health care practitioners, institutions, professional organizations, and the pharmaceutical industry to provide education about adverse drug events and their prevention. The Alerts for Patients provides useful information on herbal medicines and adverse effects. Use the search engine to locate various news items and messages concerning herbal drugs.

MedWatch: The FDA Medical Products Reporting
& Safety Information System
<http://www.fda.gov/medwatch/>

Once a drug or other medical product is approved and on the market, the U.S. Food and Drug Administration (FDA) conducts "postmarketing surveillance" to identify any safety issues as the product is used in clinical practice and, if necessary, to take appropriate action. MedWatch was created in 1993 to encourage health professionals to report details of serious drug-associated adverse reactions and other problems with medical products. MedWatch has played a major role in removing some drugs from the marketplace.

National Toxicology Program (NTP)
<http://ntp-server.niehs.nih.gov/>

The U.S. National Toxicology Program was established in 1978 by the Secretary of Health and Human Services to coordinate toxicology research and testing activities within the Department of Health and Human Services (DHHS). The NTP represents such agencies as the Food and Drug Administration, the National Institutes of Health, and the Centers for Disease Control and Prevention. NTP is currently conducting studies to identify and characterize possible adverse health effects that may be associated with prolonged use or higher doses of some of the most popular medicinal herbs, including *Ginkgo biloba, Echinacea angustifolia,* and *Panax quinquefolius* (American ginseng). The U.S. National Cancer Institute (NCI) has also recommended that the program examine the safety of aloe vera, ginseng, kava, and milk thistle to determine whether these herbs contain carcinogenic compounds.

To locate information on these studies, go the NTP home page, then select Search the NTP Study Databases. Type in the name of the herb.

Pharmaceutical Information Network
<http://pharminfo.com/>

The Pharmaceutical Information Network (PharmInfoNet) is an online drug information resource owned and operated by Mediconsult.com, Inc. This site provides access to high-quality assessments of therapeutics and to advances in new drug development, including full-text articles from clinical publications, economic data, and information from scientific symposia. A browseable Complementary Medicine section provides monographs on some of the most important herbs.

PhytoNET
<http://www.escop.com/phytonet.htm>

PhytoNET is maintained by the Center for Complementary Health Studies, University of Exeter, United Kingdom, and is designed to keep researchers up to date with the work of the European Scientific Cooperative on Phytotherapy (ESCOP). ESCOP promotes the development of therapeutic standards for herbal medicines within countries of the European Union. PhytoNET's major functions include the collection and dissemination of information on the adverse effects of herbal preparations and compiling a database of adverse reactions to herbal products reported by pharmacists and physicians. This site also provides access to the online journal *The European Phytojournal.*

The Poisonous Plant Database
  <http://vm.cfsan.fda.gov/~djw/readme.html>

This little-known bibliographic resource compiled by D. Jesse Wagstaff, DVM, provides an alphabetical listing of vascular plants known to be toxic to animals and humans. Citations are also available, listed alphabetically by author name. (Note that this is not intended to be an official FDA source.)

The Special Nutritionals Adverse Event Monitoring System (SN/AEMS)
  <http://vm.cfsan.fda.gov/~dms/aems.html>

The Special Nutritionals Adverse Event Monitoring System is a searchable database reporting adverse effects and problems associated with the use of special nutritional products, such as dietary supplements (including herbs), infant formulas, and medical foods. Data, which date back to 1993, are obtained from MedWatch, through an FDA District or Field Office, from health professionals, and from other sources. Search the SN/AEMS database if you have a question about a specific herb and whether an adverse event has occurred with its use. Results are returned in a table format, showing each product's name, reported adverse effects, name of manufacturer, and all ingredients.

It is important to remember that this is a database of "reported" adverse reactions. The absence of a report does not necessarily mean a particular product or ingredient is not associated with an adverse reaction.

TOXLINE
  <http://igm.nlm.nih.gov/>

TOXLINE contains citations from 1965 to the present and covers information on the pharmacological, biochemical, physiological, and toxicological effects of drugs, including herbs. As of this writing, this database is accessible through the NLM's Internet Grateful Med (IGM) System. Future plans include the replacement of IGM by the new NLM Gateway <http://gateway.nlm.nih/gw/Cmd>.

Chapter 11

Quackery, Health Fraud, and Consumer Protection

There are two ways to slide easily through life: to believe everything or to doubt everything. Both ways save us from thinking.

Alfred Korzybski

CONFLICTS BETWEEN ALLOPATHIC MEDICINE AND HERBALISTS

Health professionals most frequently criticize herbal practitioners for what they see as a lack of scientific evidence indicating that herbal remedies are effective. They are also disturbed by what they view as an entrenched "antiscientific" attitude among many practicing herbalists. To someone trained in the methodology of Western scientific medicine, some of the writing about herbs can seem decidedly "flaky," with an aura of mysticism and mythology and links to the New Age movement, with its beliefs in astrology, crystals, and Tarot cards. Health professionals also object to some of the basic tenets of traditional herbalism, such as the belief in the superiority of herbs over conventional drugs, the "activation" of the body's healing system by herbs, a belief in the "synergistic," or combined, activities of several herbal constituents, and an insistence that knowledge handed down from generation to generation can be as valid as that gleaned from clinical trials.[1,2]

From their point of view, traditional herbalists believe that allopathic medical practitioners do not understand the holistic nature of herbal practice, with its emphasis on the prevention of illness and treatment of the whole person. In the words of one prominent herbalist, holistic medicine is

> . . . an appreciation of patients as mental and emotional, social and spiritual, as well as physical beings. It respects their capacity for healing themselves and regards them as active partners in, rather than passive recipients of, health care.[1]

Herbalists also criticize the current approach to determining the efficacy of herbal remedies, in particular clinical trials that focus on a single purified ingredient. One of the basic tenets of herbalism is that while standard drugs are single chemical entities, herbs may contain several or many compounds, which work in synergy, and it is the activity of the plant itself, not its chemical constituents, that is most important.[1]

Many allopathic practitioners, with their training in scientific medicine, regard many CAM therapies with a great deal of skepticism, believing it to be a passing fad, a product of Western society's current dissatisfaction with science. Others remain more open-minded but express concern because of the opportunities for fraud and quackery, especially on the Internet. Many patients with life-threatening diseases or conditions, such as cancer and HIV/AIDS, seek out experimental treatments that not only are unproven but can be dangerous. For example, in the 1970s, a substance called "laetrile," isolated from apricot pits, was heavily promoted as a natural anticancer compound. Despite its possible harmful effects, and a lack of any evidence that it does have anticancer properties, it is now being sold by companies via the Internet.[3]

HEALTH FRAUD AND QUACKERY

Due to the current absence of an effective regulatory framework in the United States, it is often difficult for the health practitioner or consumer to identify the serious science from the quackery. The growth of the Internet has exacerbated the problem. Web sites and bulletin boards abound with exotic, unproven, or exaggerated claims, and many opportunities exist for the unscrupulous to promote useless or harmful remedies to an unsuspecting public. The Federal Trade Commission (FTC) has estimated that American consumers waste billions of dollars on unproven, fraudulently marketed, and sometimes useless health care products and treatments.[4] After a campaign to identify health fraud on the Internet, the FTC concluded that many Web sites make deceptive, unproven, and fraudulent health claims.[3]

The American physicist Richard Feynmann once remarked, "It's good to have an open mind, but not so open that your brains fall out." This chapter therefore includes some Web sites that are very skeptical of the value of herbal therapies. Their inclusion is by no means an endorsement of their views; rather, it is an attempt to maintain some sort of balance and to alert readers to writings by both advocates and critics of herbalism. Other included sites are primarily concerned with the way the Internet is opening up new opportunities for health fraud and quackery, and not just in the CAM area. Note the subtle distinction between health fraud and quackery: the FDA describes health fraud as "the promotion, advertisement, distribution, or sale of articles, intended for human or animal use, that are represented as being effective to diagnose, prevent, cure, treat, or mitigate disease (or other conditions), or provide a beneficial effect on health, but which have not been scientifically proven safe and effective for such purposes." The word quackery has a slightly different meaning; *The American Heritage Dictionary* defines quack as "an untrained person who pretends to have medical knowledge."

WEB SITES DEDICATED TO CONSUMER EDUCATION AND AWARENESS

American Council on Science and Health (ACSH) <http://www.acsh.org/>

The American Council on Science and Health (ACSH) is a nonprofit organization founded by a group of scientists who were concerned that "many important public policies related to health and the environment did not have a sound scientific basis." It is dedicated to the education of consumers on issues relating to food, nutrition, chemicals, pharmaceuticals, lifestyle, the environment, and health.

The American Public Health Association (APHA) <http://www.apha.org/>

The American Public Health Association (APHA) is the oldest and largest organization of public health professionals in the world, representing more than 50,000 members from over fifty occupations of public health. Go to the Legislative Issues & Advocacy section for the APHA's policy statements on alternative medicine/therapy. Use the site's search engine to locate many other items concerning alternative medicine and public health.

Center for Medical Consumers (CMC)
<http://www.medicalconsumers.org/>

The CMC, a nonprofit advocacy organization based in New York, is active in both statewide and national efforts to improve the quality and availability of health care. It is specifically committed to educating the public about issues affecting the quality of the American health care system, including critiques of drug advertisements. New items and newsletter articles pinpoint CAM issues and therapies.

Center for Science in the Public Interest (CSPI)
<http://www.cspinet.org/>

The CSPI was founded in 1971 by scientists who had previously worked for Ralph Nader's Center for the Study of Responsive Law. The center is a nonprofit education and advocacy organization that focuses on improving the safety and nutritional quality of the food supply. It represents citizens' interests before legislative, regulatory, and judicial bodies and was involved with petitioning the FDA about labeling requirements for dietary supplements. Go to the Nutrition Action Healthletter link for its Health Watch section, with feature articles and news items assessing current information on the efficacy of herbal remedies.

ConsumerLab.com
<http://www.consumerlab.com>

ConsumerLab.com (CL) is a new, independent research laboratory that investigates the product quality of commercially available dietary supplements, including herbal products. ConsumerLab.com aims to test natural products to see if they meet scientific standards for potency, purity, and consistency. Products that pass CL's testing are eligible to bear the CL Seal of Approval. Results from its analyses of several popular herbal products, including Asian and American ginseng, saw palmetto, and ginkgo biloba, are available from the Web site.

A Dictionary of Alternative-Medicine Methods
<http://www.acsh.org/dictionary/index.html>

This online book is a comprehensive dictionary of alternative medicine methods, ranging from "Bach flower therapy" to "Zulu Sangoma bones." The dictionary is prepared by Jack Raso, MS, RD, Director of Publications at the American Council on Science and Health (ACSH) (see the ACSH entry in this chapter). It is far from being an unbiased resource, as the author admits, but it is a useful guide nevertheless.

The Federal Trade Commission (FTC)
<http://www.ftc.gov/>

The FTC Bureau of Consumer Protection
<http://www.ftc.gov/ftc/consumer.htm>

The FTC is the U.S. government agency responsible for monitoring business practices in the United States. Its duties include enforcing antitrust laws to ensure that competition is fair, preventing false and deceptive advertising, regulating labeling and packaging, and gathering data concerning business conditions. In addition to its enforcement responsibilities, it is also mandated to advance congressional policies through cost-effective nonenforcement activities, such as consumer education. The FDA and the FTC share jurisdiction over the regulation of dietary supplements; whereas the FDA oversees labeling requirements, the FTC regulates other aspects of advertising and has the power to order a business to cease and desist using an advertisement found to be deceptive.

According to the market research company Cyber Dialogue, as of December 1999, 34.7 million U.S. adults went online to find health information, an increase of 56 percent over the previous year, and this is projected to increase to 88.5 million adults by 2005.[5] Consumers not only are turning to the Internet for health information but are using it as an online pharmacy. The FDA is very concerned about the public health implications of the sale of unapproved drugs on the Internet, since this practice can bypass the usual public health safeguards set in place by Congress. Electronic advertising, similar to its print counterparts, is subject to regulation by the FTC, and in recent years the agency has increased its efforts to deal with Internet health sites promoting dubious health products. After a Health Claim Surf Day in 1997, and another in 1998, FTC's investigators were able to identify 800 Web sites making fraudulent health claims.[6] In 1999, it launched Operation Cure.All, a campaign intended to stop companies from making deceptive claims about health products, and, as a result, four companies were charged with making unsubstantiated health claims for products advertised on the Internet.[8] Go to the Consumer Protection section of the FTC Web site for full-text articles on advertising policy and regulations, as well as guidelines for identifying health fraud.

**Georgia Council Against Health Fraud's Healthcare
Reality Check
  <http://www.hcrc.org/>**

The Georgia Council Against Health Fraud (GCAHF), a nonprofit volunteer organization affiliated with the National Council Against Health Fraud (see the NCAHF entry in this chapter), seeks to provide science-based information on alternative and complementary medicine. Healthcare Reality Check (HCRC) provides a variety of useful resources, news items about the misuse of CAM therapies, and an HCRC encyclopedia with evaluative information on CAM methods and providers.

**InteractionReport.org
  <http://www.interactionreport.com/>**

This is a useful resource for information on drug-herb interactions and adverse reactions involving herbs and nutritional supplements. Information is provided by a consortium of educational institutions, professional associations, hospitals, insurance companies, managed care organizations, and members of the pharmaceutical, supplement, and herbal products manufacturing and distribution industry. Members include the American Herbalists Guild, the American Association of Naturopathic Physicians, and the Southwest College of Naturopathic Medicine.

The National Council Against Health Fraud (NCAHF)
<http://www.ncahf.org/>

The NCAHF is a nonprofit, voluntary health agency that focuses on health fraud, misinformation, and quackery as public health problems. The organization is composed of health professionals, educators, researchers, attorneys, and concerned citizens. Useful resources include the "NCAHF Position Paper on Over-the-Counter Herbal Remedies (1995)" and its "Consumer Health Digest (CHD)," a free weekly e-mail newsletter summarizing scientific reports, legislative developments, and enforcement actions. CHD also contains Web site evaluations and book critiques. The Webmaster is listed as Stephen Barrett, MD, the founder of the Quackwatch Web site (see the corresponding entry in this chapter).

NutriWatch
<http://www.nutriwatch.org/>

Subtitled "Your Guide to Sensible Nutrition," this site was launched in February 2000 as part of the Quackwatch system and is operated by Stephen Barrett, MD (see the following entry). It aims to provide comprehensive information about nutrition and food safety issues.

Quackwatch
<http://www.quackwatch.com/>

Quackwatch, subtitled "Your Guide to Health Fraud, Quackery, and Intelligent Decisions," is a unique Web site dedicated to the exposure of what the site's authors view as "health-related frauds, myths, fads, and fallacies." It includes many critiques of CAM therapies and procedures. Quackwatch is operated by Stephen Barrett, MD, the medical editor of Prometheus Books, author of several books on questionable medical practices, and consulting editor to *Nutrition Forum,* a newsletter emphasizing the exposure of fads, fallacies, and quackery. This is a large site, with an extensive collection of resources. Go to the Herbal Practices and Products and Dietary Supplements, Herbs, and Hormones sections for a large collection of relevant online articles. The site also features a useful bibliography of consumer protection publications and much valuable information on the commercial interests behind some of the more popular consumer-oriented herbal Web sites.

The Skeptic's Dictionary: A Guide for the New Millennium
<http://skepdic.com/>

The Skeptic's Dictionary: A Guide for the New Millennium provides definitions, arguments, and essays on such subjects as the supernatural, occult, paranormal, and pseudoscience.

Chapter 12

Information for Consumers and Patients

The next major advances in the health of the American people will be determined by what the individual is willing to do for himself and for society at large.

John H. Knowles
Doing Better and Feeling Worse:
*Health in the United States,*1977

THE INTERNET AND HEALTH CARE

The rise of the managed care movement in health care has resulted in a heightened focus on cost-effective health and disease prevention. At the same time, the rise of the Internet has opened up a vast reservoir of health and medical information to any individual with a computer and an Internet access account, and an increasing number of people are seeking information and advice from sources other than their primary physicians. The next decade may witness a revolution in health care fueled by computer-literate consumers exploiting the communication possibilities of the Internet. Already the Internet is being used, not only to access a plethora of health-related information and to communicate with health care providers, but as a means to create networks of knowledgeable consumers who wish to take an active role in their own health care.

As the number of Internet users continues to mushroom, many physicians report a growing number of patients asking for recommended online health sites. Many patients already expect to be able to communicate with their doctor through electronic mail.[1] A new type of patient is emerging: someone who no longer just accepts what he or she is told by the physician but who carries out extensive research on the Internet to learn about his or her condition and the treatment options available. The Internet will allow the growth of online communities seeking to play a major role in improving or maintaining their own health, or in recovering from diseases or injuries. According to Dr. Tom Ferguson, a pioneer in consumer health information, "self-help communities in cyberspace" will become increasingly important as more and more patients turn to the Internet for information and support.[1]

The past decade has seen a significant increase in the American public's general awareness and use of CAM. Interest in herbal medicine seems greater than ever, and herbal products continue to gain in popularity. This visibility, coupled with the large quantities of medical information on the Internet, has given many people the increased confidence to attempt their own treatment for a variety of ailments. As discussed elsewhere, some online herbal information is misleading, biased, or promoted by the manufacturers of herbal products. Health professionals are justifiably concerned that consumers are not being given adequate information about the use of herbs and their possible dangers. Recognizing that this is a problem, federal health agencies, university medical centers, consumer advocacy groups, and other respected groups or individuals have established online gateways guiding consumers to the most reliable and unbiased information.

WEB SITES SERVING AS GATEWAYS
TO RELIABLE HERBAL INFORMATION

AMA Health Insight
  <http://www.ama-assn.org/ama/pub/category/>

Follow the link for Patients, then Health Information. This site, a joint venture of the American Medical Association (AMA) and Loma Linda University, provides peer-reviewed information for patients about specific diseases and conditions.

Ask Dr. James Duke
  <http://www.herbalvillage.com/index6.html>

Dr. James Duke is an internationally known ethnobotanist, having worked for the Missouri Botanical Garden and the U.S. Department of Agriculture's Medicinal Plant Laboratory. He is the author of both scholarly and popular books on medicinal herbs and has developed several important databases for information about herbs. (See Chapter 7.)

Ask Dr. Weil
<http://www.drweil.com/>

Andrew Weil, MD, is arguably the best-known and one of the most influential physician-writers in the area of CAM therapies in the United States today. A strong advocate of integrative medicine, the integration of traditional and CAM remedies, he is currently Director of the Program in Integrative Medicine and Clinical Professor of Internal Medicine at the University of Arizona in Tucson. (See the entry for University of Arizona Program in Integrative Medicine in Chapter 6.)

Cochrane Collaboration Consumer Network
<http://www.cochraneconsumer.com/>

These are synopses of Cochrane systematic reviews written for consumers, arranged in alphabetical order of review title (see the entry for The Cochrane Collaboration in Chapter 9). These summaries are provided by the Health Information Research Unit (HIRU) at McMaster University in Canada.

Consumer.gov
<http://www.consumer.gov/>

This is the U.S. government's consumer information gateway to federal government resources on food, product safety, health, financial services, and other consumer topics.

Consumer Reports Online
<http://www.consumerreports.org/>

Consumer Reports is a well-known consumer protection magazine seeking to provide unbiased and accurate information on items ranging from computers to herbal drugs. These reports are designed to assist consumers in making more informed decisions about their personal health care. While a paid subscription is required to view the complete text of this publication, selected articles may be accessed for free, and some sections of the Web site have useful health-related information, such as a consumer guide to medical research on the Internet.

FDA Consumer **Magazine**
<http://www.fda.gov/fdac/default.htm>

FDA Consumer is an official FDA magazine reporting on FDA activities and providing in-depth health information for the consumer. It is one of the most important Internet sites for accurate consumer information on herbs and other dietary supplements. For example, searching on "herbs" finds numerous articles and news items, including the following documents: *An FDA Guide to Dietary Supplements, Harvesting Drugs from Plants,* and *Dietary Supplements Associated with Illnesses and Injuries.* Online full text is available going back to July-August 1995. Access the magazine using the URL given here, or use the link on the FDA's home page at <http://www.fda.gov/>.

Go Ask Alice!
<http://www.goaskalice.columbia.edu/>

Go Ask Alice! was created by the Health Education Program at New York's Columbia University. It is a health question-and-answer Internet site designed to provide "factual, in-depth, straightforward and nonjudgmental information to assist readers' decision-making about their physical, sexual, emotional and spiritual health." Users can post questions anonymously and have them answered by university health educators and practitioners.

Good Housekeeping **Magazine**
<http://goodhousekeeping.com/>

Good Housekeeping is known for providing a quality seal to consumer products, ranging from dishwashers to clothing, and has recently ventured into the area of complementary and alternative medicine. The Good Housekeeping Seal of Approval has been alerting American consumers as to product quality for almost a century. Companies are allowed to put the seal on their products following a review by the Good Housekeeping Institute. It is customary for the magazine to test products before accepting advertising, and it now requires dietary supplement companies who want to advertise in the publication to meet a standard for product purity.

For consumer information on herbs and drugs interactions, select Health Articles from the home page.

healthfinder
<http://healthfinder.gov/>

Developed by the U.S. Department of Health and Human Services (DHHS) in collaboration with other federal agencies, healthfinder is a free gateway to reliable consumer health-related information sources. It includes links to selected online publications, databases, Web sites, and support and self-help groups. The home page has a "hot topics" section: follow the "alternative medicine" link to access "alternative therapies" and "herbal resources."

HealthWorld Online
<http://www.healthy.net/>

This Web site is aimed at the educated consumer wishing to be more involved in the management of his or her own health care and treatment. It describes itself as "a virtual health village where you can access information, products, and services to help create your wellness-based lifestyle," by supplying information and providing an online medium for interactive social support. At the home page click on the Alternative Therapies link to access a comprehensive collection of CAM resources, with items on Western herbal medicine, traditional Chinese medicine (TCM), and Ayurvedic medicine. Of particular note is the Herbal Medicine subsection, with articles by David Hoffman, BSc, MNIMH, a noted herbalist and author of several respected books on herbs.[2]

IBIDS (International Bibliographic Information
on Dietary Supplements) Database
<http://ods.od.nih.gov/databases/ibids.html>

Created by the Federal Office of Dietary Supplements (ODS), the IBIDS database is an online bibliographic resource of over 380,000 scientific citations and abstracts, designed to provide ready access to the international scientific literature on vitamins, minerals, hormonal products, and botanicals. IBIDS now features a new file, IBIDS Consumer Database, containing consumer-oriented articles. (For more information, see the IBIDS entry in Chapter 13.)

The Longwood Herbal Task Force (LHTF)
<http://www.mcp.edu/herbal/default.htm>

The LHTF was founded in the fall of 1998 by faculty, staff, and students from Children's Hospital, the Massachusetts College of Pharmacy and Health Sciences, and the Dana Farber Cancer Institute. Its mission is to help educate clinicians about the use of herbal remedies and other dietary supplements. The LHTF has comprehensive collections of fact sheets on herbs for both patients and their health care providers. Documents are available for the more popular herbs, such as echinacea and St. John's wort, but also for lesser-known herbs, such as rhubarb *(Rheum officinale),* dandelion *(Taraxacum officinalis),* and bearberry *(Uva ursi).* Information on herb-drug and herb-nutrient interactions is also available.

MayoClinic.com
<http://www.mayohealth.org/>

The Mayo Clinic, in Rochester, New York, is one of the most respected medical institutions in the world. Relevant online consumer resources include the *United States Pharmacopeia Drug Guide,* a searchable database of information on over 8,000 prescription and over-the-counter medications, listed by brand name; an Ask the Dietitian section, which has a wealth of information on vitamins, minerals, and supplements; and an Alternative Medicine section, with consumer-oriented articles on herbal supplements.

MEDLINE*plus*
<http://www.nlm.nih.gov/medlineplus/>

MEDLINE*plus* CAM Information
<http://www.nlm.nih.gov/medlineplus/alternativemedicine.html>

MEDLINE*plus* is a Web-based information service from the National Library of Medicine (NLM). It is designed "to facilitate consumer access to medical information resources, providing links to full-text resources, dictionaries, organizations, directories, libraries, and clearinghouses for answers to health questions." Note that each health topic is linked to a preformulated MEDLINE search created by medical librarians, allowing the user to retrieve an up-to-date list of relevant citations on specific topics from the PubMed database. Click on Health Topics, then select Herbal Medicine from the List of All Topics to access herbal resources.

National Health Information Center (NHIC)
<http://www.health.gov/nhic/>

The NHIC is a health information referral service established in 1979 by the Office of Disease Prevention and Health Promotion (ODPHP), U.S. Department of Health and Human Services (DHHS), to provide free access to the Health Information Resource Database (HIRD) publications and a number of referral publications and policy documents. The database includes information on over 1,100 organizations and government offices that provide health information upon request. Entries include contact information, abstracts, and information about publications and services.

The Natural Pharmacist
<http://www.tnp.com/>

This site provides referenced information on various herbs and other supplements for the consumer and health professional. It includes a comprehensive drug-herb interaction section, a disease-specific page, and online discussion forums for conditions such as migraines and arthritis.

NOAH: New York Online Access to Health
<http://www.noah-health.org/index.html>

New York Online Access to Health (NOAH) is a bilingual (English and Spanish) consumer health information site. NOAH is a collaborative project between The City University of New York, The Metropolitan New York Library Council, The New York Academy of Medicine, and The New York Public Library. It is particularly focused on the underserved population of health consumers, many of whom are Spanish speaking.

Reuters Health eLine
  <http://www.reutershealth.com/frame2/eline.html>

Reuters is a leading provider of general and business information from around the world. Reuters Health eLine is a consumer-oriented medical news service providing the top clinically relevant health care news stories. Approximately fifteen new items are added each day, and there is no charge to view those which are less than thirty days old. From the drop-down boxes, select the topic Alternative Medicine, then choose a date.

Women.com
  <http://www.women.com/>

Women.com is an Internet network dedicated to women's issues and information. It features original articles and personalized services, as well as community and online shopping. The network is composed of more than 90,000 pages of content organized into twenty topical "channels," including career, entertainment, family, health, home, horoscopes, technology and the Internet, and pregnancy. Women.com also features selected online content from some of the world's leading women's magazines. Registration is required to access all sections of this site, including news groups and chat rooms.

Chapter 13

MEDLINE and Other Online Indexes/Databases

We are drowning in information and starving for knowledge.

Rutherford D. Roger

MEDICAL INFORMATION
AND THE INFORMATION EXPLOSION

Rapid technological advances, including the growth of computer networks and the World Wide Web, have had a significant impact on the amount of biomedical information now being produced. Back in the 1960s and 1970s, the idea of an "information explosion" began to gain ground when studies by Derek J. de Solla Price showed that the scientific literature was doubling every fifteen years.[1] In 1994, Humphreys and McCutcheon looked at growth patterns in the National Library of Medicine (NLM)'s serials collection and its *Index Medicus* journals and came to the same conclusion for the medical literature.[2] It is now estimated that in the biomedical sciences more than 2 million journal articles are published each year.[3]

Rapid advances in computer technology have revolutionized many areas of the sciences and further accelerated the rate at which information is being produced. According to IBM, areas such as pharmaceutical research and the Human Genome Project (HGP) are producing such vast quantities of data that the amount of information in the life sciences is now expected to double every six months.[4] The "information age," combining rapidly developing information technology and massive growth in biomedical and clinical data, is placing special demands on health care workers who are finding it increasingly difficult to keep abreast of new therapies and procedures.

THE LITERATURE ON CAM

Practically all health professionals now have some familiarity with the MEDLINE database, a tool that has evolved as a "filter" for the journal literature. In the United States, and in many other countries, it has become the primary database for searching the mainstream biomedical research literature, and it would be nearly impossible to locate relevant published journal articles without it. However, when using MEDLINE to locate information on herbal medicine and other CAM areas, three things need to be remembered. First, MEDLINE is intentionally selective in what it includes, indexing less than half of the estimated 10,000 medical journals published each year. Thus, a good deal of information that is published in journals is, for one reason or another, not represented in the MEDLINE database. Second, information related to the safety and efficacy of CAM interventions is difficult to locate and is widely scattered throughout the literature. It often is found in the so-called gray literature, such as trade journals, pamphlets, conference proceedings, and market research reports, and therefore difficult to identify.[5] According to a recent study by the National Library of Medicine, CAM information can be found in nearly 700 journals published by many different countries, and in more than 150 electronic databases.[6] Finally, a significant number of CAM studies are first initially published in languages other than English.[7] For example, some of the most important data on herbal drugs is to be found in the German pharmaceutical literature, the bulk of which needs to be translated and is not indexed by American indexing services. A search for comprehensive information will therefore have to include bibliographic databases other than MEDLINE that have stronger collections in the European pharmaceutical literature. For example, Elsevier Science's EMBASE has more comprehensive international coverage than MEDLINE and is thus stronger in many areas of CAM.

Government-funded centers, such as the National Center for Complementary and Alternative Medicine (NCCAM) and the Office of Dietary Supplements (ODS), have begun to tackle some of the problems mentioned by producing new online information resources specifically designed for retrieving CAM information, such as the IBIDS database and CAM on PubMed. Other organizations, such as the Southwest School of Botanical Medicine, have made available electronic versions of their own in-house collection of specialized bibliographic resources. More specialized nonbibliographic factual databases can now also be easily accessed on the Web, making available hard-to-find data such as chemical constituents and medicinal uses of medicinal plants as part of traditional healing systems.

WEB SITES PROVIDING ACCESS TO DATABASES AND INDEXES

Africa Journals Online (AJOL)
<http://www.inasp.org.uk/ajol/index.html >

AJOL provides access to African journals in the agricultural sciences, science and technology, as well as health and social sciences. For details, see the corresponding entry in Chapter 7.

AfricaBib
<http://www.africabib.org/>

AfricaBib contains the Bibliography of Africana Periodical Literature Database and the Bibliographic African Women's Database. For details, see the corresponding entry in Chapter 7.

AGRICOLA (AGRICultural OnLine Access)
<http://www.nal.usda.gov/ag98/>

National Agricultural Library (NAL) Homepage
<http://www.nal.usda.gov>

AGRICOLA is produced by the U.S. National Agricultural Library (NAL), in cooperation with certain land-grant universities and research institutions. It covers all aspects of agriculture and allied disciplines, such as natural resources and nutrition. Though it is primarily useful for information on the cultivation of herbs and other medicinal plants, it does contain a surprisingly large number of unique citations to articles on other herb-related topics. Of particular importance for the herbal community is its indexing of *HerbalGram*, the American Botanical Council's peer-reviewed herbal journal.

The database is organized into two bibliographical data sets, which can be searched separately: an Online Public Access Catalog, known as "Books, etc.," for information on the library's extensive collection of books serials, audiovisuals, and other resources held by the NAL, and the Journal Article Citation Index, known as "Articles, etc.," for citations of journal articles (many with abstracts), book chapters, short reports, and reprints. The NAL home page freatures a prominent link to AGRICOLA.

Note that selected AGRICOLA entries are included in the Office of Dietary Supplement's International Bibliographic Information on Dietary Supplements (IBIDS) database (see the entry for IBIDS in this chapter).

AIDSLINE
<http://igm.nlm.nih.gov/>

Many people with HIV/AIDS have turned to CAM therapies to manage some of their health problems. Several herbal formulations are currently being investigated, including Ayurvedic and Chinese herbal formulations. AIDSLINE (AIDS Information Online), produced by the National Library of Medicine, is a collection of bibliographic citations focusing on research, clinical aspects, and health policy issues related to AIDS. This database includes articles and other documents, such as government reports, technical reports, meeting abstracts and papers, monographs, special publications, theses, books, and audiovisual materials that are not included in MEDLINE.

As of this writing, this database is accessible through the NLM's Internet Grateful Med (IGM) system. Future plans include the incorporation of journal records into the PubMed system (see PubMed entry in this chapter) and the eventual replacement of IGM by a new NLM Gateway <http://gateway.nlm.nih.gov/gw/Cmd>.

AMED: Allied and Complementary Medicine Database
<http://www.bl.uk/services/stb/amed.html>

Even though this is not a free online database, information about the AMED database is included because it is one of the most important resources for CAM information. The database, produced by the Health Care Information Service of the British Library, contains bibliographic citations with abstracts, covering subjects such as acupuncture, osteopathy, Chinese medicine, homeopathy, rehabilitation, occupational therapy, hypnosis, herbalism, physiotherapy, chiropractic, holistic treatments, and other professions allied to medicine. Many of its journals are not indexed by other biomedical sources, and though the coverage is strongly European, the majority of titles are in English. Fee-based searching via the Internet is provided by several commercial database vendors.

BIOETHICSLINE
<http://igm.nlm.nih.gov/>

Bioethics Thesaurus
<http://www.georgetown.edu/research/nrcbl/ir/biothes.htm>

BIOETHICSLINE indexes literature related to the ethical and public policy aspects of health care and biomedical research. The database is produced by the National Library of Medicine (NLM), the National Human Genome Research Institute, and the Bioethics Information Retrieval Project at the Kennedy Institute of Ethics, Georgetown University. As of this writing, this database is accessible through the NLM's Internet Grateful Med (IGM) system. Future plans include the incorporation of journal records into the PubMed system (see PubMed entry in this chapter) and the eventual replacement of IGM by a new NLM Gateway <http://gateway.nlm.nih.gov/gw/Cmd>. Specific CAM and herb articles in this database relate to questions involving the provision of CAM information by health professionals and controversial areas of herbal medicine, such as herbal-induced abortion and contraception. Articles in BIOETHICSLINE are indexed using the *Bioethics Thesaurus,* an indexing vocabulary of approximately 700 terms designed for this database.

CAM on PubMed
<http://www.nlm.nih.gov/nccam/camonpubmed.html>

CAM on PubMed was developed jointly by the National Library of Medicine (NLM) and the National Center for Complementary and Alternative Medicine (NCCAM). Since the direct URL is quite long and complex, it is best accessed by using the link from the NCCAM's Web site. (You can also search directly from PubMed, but you need to limit the search to the CAM subset: to do this, click on the "Limit" function, then select the "Complementary Medicine" Subset from the pull down menu.)

Cam on PubMed is basically a subset of citations from the National Library of Medicine's MEDLINE database, extracted using Medical Subject Headings (MeSH), primarily in the "alternative medicine" section of the MeSH tree structure (see Figure 13.1.) (See also the entry for PubMed for a more detailed description of MeSH.) The database consists of approximately 180,000 bibliographical citations, dated from 1966 to the present. Though the experienced searcher can probably get the same results using the parent MEDLINE database, a database of approximately 220,000 citations is obviously easier to deal with than the 11 million in MEDLINE.

CANCERLIT
<http://cancernet.nci.nih.gov/cancerlit.shtml>

CANCERLIT is a bibliographic database produced by the National Cancer Institute's International Cancer Information Center. It contains more than 1.5 million citations and abstracts from over 4,000 different sources on all aspects of cancer. The database may be searched using MeSH terms (see the PubMed entry in this chapter).

FIGURE 13.1. Part of the Alternative Medicine Section of the MeSH Tree Structure

```
Homeopathy
Imagery (Psychotherapy)
Kinesiology, Applied
Massage
        Acupressure
Medicine, Traditional
        Medicine, African Traditional
        Medicine, Arabic
                Medicine, Unani
        Medicine, Ayurvedic
        Medicine, Herbal
                Medicine, Kampo
                Medicine, Unani
        Medicine, Oriental Traditional
                Medicine, Chinese Traditional
                Medicine, Kampo
        Shamanism
Mental Healing
Mind-Body Relations (Metaphysics)
Moxibustion
Music Therapy
Naturopathy
```

CancerNet—Physician Data Query (PDQ)
<http://cancernet.nci.nih.gov/pdq.html>

PDQ (Physician Data Query) is a National Cancer Institute database that contains the latest peer-reviewed information about cancer treatment, screening, prevention, genetics, and supportive care, plus clinical trials. It consists of three main files: a Cancer Information File, covering prognostic and treatment information for over seventy-five types of cancer; a Protocol File, with information on active or approved clinical trials and a limited number of standard treatment regimens; and a Directory File, which provides an international register of names, addresses, and telephone numbers of clinicians and institutions devoted to the care of cancer patients.

Go to the Treatment section for its Complementary and Alternative Medicine Summaries document for mistletoe (*Viscum album* L.). This CAM information summary provides an overview of the use of mistletoe as a treatment for cancer, including a history of mistletoe research, results of clinical trials, and possible side effects of mistletoe use.

The Centralized Information Service for Complementary Medicine (CISCOM)
<http://www.rccm.org.uk/cisc.htm>

The CISCOM database is produced by the Research Council for Complementary Medicine (RCCM) in the United Kingdom. It has information on over 4,000 randomized trials and over 60,000 citations and abstracts covering all the major CAM therapies.

Combined Health Information Database (CHID)
<http://chid.nih.gov/>

CHID is a bibliographic database produced by federal health-related agencies, including the Centers for Disease Control (CDC), the Office of Disease Prevention and Health Promotion, the Department of Veterans Affairs, and the National Center for Complementary and Alternative Medicine (NCCAM). The database provides bibliographic citations for major health journals, books, reports, pamphlets, audiovisuals, hard-to-find information resources, and health education/promotion programs. At present, CHID covers seventeen topics, in seventeen subfiles, including a Complementary and Alternative Medicine (AM) section. The types of documents in the subfiles are quite broad, and CHID includes a wealth of health promotion and consumer-oriented education materials and program descriptions that are not indexed elsewhere, including newsletters, audiovisual material, cookbooks, coloring books, and cartoon books.

CRISP: Computer Retrieval of Information on Scientific Projects
<https://www-commons.cit.nih.gov/crisp/>

CRISP is a database of biomedical research projects conducted at universities, hospitals, and other research institutions funded by the U.S. Public Health Services. The database is maintained by the Office of Extramural Research at the National Institutes of Health (NIH). It includes projects funded by the Food and Drug Administration (FDA), Centers for Disease Control and Prevention (CDCP), NIH Substance Abuse and Mental Health Services (SAMHSA), Health Resources and Services Administration (HRSA), Agency for Healthcare Research and Quality (AHRQ), and Office of Assistant Secretary of Health (OASH). To search, select Query Form from the home page. CRISP can be searched by project title, principal investigator's abstract, principal investigator, or grant-receiving institution. Users can employ terms from the CRISP thesaurus for more precise retrieval—this is not available online, but terms can be viewed in relevant records then used for searching.

The Datadiwan
<http://www.datadiwan.de/>

The most accurate and comprehensive information on the efficacy of herbs is to be found in the German pharmaceutical literature. Unfortunately, many cited references for herbs are written in German and published in journals not indexed by MEDLINE. The Datadiwan Web site from Germany is a complex network of databases: it describes itself as a database to "holistic medicine and frontier [sic] sciences," as well as "a scientific discussion forum" and "a network which links research institutions and organizations." The Patienteninformation für Naturheilkunde (patient information for natural therapies) includes a searchable database providing access to over 5,000 bibliographic citations. Many of the documents are in German, but help for using the site is in English: select English Version from the bottom of the home page.

(The name Datadiwan is formed from the words *data* and *diwan*. The second corresponds to the English term *divan*—a long seat much like a sofa! It is a metaphor for making access to data more comfortable for the user.)

Dr. Felix's Free MEDLINE Page
<http://www.docnet.org.uk/drfelix/>

If you do not wish to use the NLM's free PubMed system, several other Internet sites provide free access to the MEDLINE database. This site provides useful information about these alternate sites.

EthnobotDB—Worldwide Plant Uses
<http://ars-genome.cornell.edu/Botany/aboutethnobotdb. html>

This resource was compiled by Dr. James A. Duke and Stephen M. Beckstrom-Sternberg of the USDA. EthnobotDB provides information concerning worldwide medicinal use of over 80,000 plants. The database may be searched by keyword or browsed by common name, country, taxonomic class (i.e., by genus, family), or use.

esp@cenet: Europe's Network of Patent Databases
<http://ep.espacenet.com/>

The British Library estimates that 80 percent of the original information published in patents is never published in the journal literature. The patent literature is a much neglected source of information on drug development. With the current resurgence of interest in herbal preparations, many pharmaceutical companies are seeking ways to obtain patents for medicinal preparations containing herbal compounds. Each issue of the American Botanical Council's *HerbalGram* now includes synopses of issued plant patents.

Free searchable databases, with links to full text, are now available for patents issued by the European Patent Office (EPO) and the U.S. Patent and Trademark Office (USPTO). Web sites for Russian and Japanese patents have also recently become available.

European Patent Office (EPO)
<http://www.european-patent-office.org/index.htm>

HISTLINE (History of Medicine Online)
<http://igm.nlm.nih.gov/>

The National Library of Medicine's bibliographic database covers literature about the history of health-related professions, sciences, individuals, institutions, drugs, and diseases in all parts of the world and all historic periods. (For further information, see the corresponding entry in Chapter 2.)

IBIDS (International Bibliographic Information
on Dietary Supplements) Database
<http://ods.od.nih.gov/databases/ibids.html>

In January 1999, the ODS launched the International Bibliographic Information on Dietary Supplements (IBIDS) database, an online bibliographic resource of over 380,000 scientific citations and abstracts, designed to provide ready access to the international scientific literature on vitamins, minerals, hormonal products, and botanicals. IBIDS was codeveloped with the Food and Nutrition Information Center, National Agricultural Library, and U.S. Department of Agriculture. The database is updated quarterly.

The first phase of development for IBIDS involves adding information about dietary supplements from the peer-reviewed scientific literature, from 1985 to the present. The second phase (not yet begun) is scheduled to involve the addition of monographs, government reports, and symposia proceedings. IBIDS currently contains citations from three major databases: the National Library of Medicine's MEDLINE database, the National Agricultural Library's AGRICOLA (AGRIcultural OnLine Access) database, and AGRIS (International Information System for the Agricultural Sciences and Technology), produced by the Food and Agriculture Organization (FAO) of the United Nations. The goal is to increase the comprehensiveness of the database in the future by incorporating citations from other relevant food technology and biological databases.

The American Herbal Products Association (AHPA) estimates that, in the United States, over 550 herbal ingredients are available in a wide variety of "supplement" products. However, for the time being, herbal citations in IBIDS are limited to what it considers to be the eighty-five most widely used herbs or "key species." This list is partially based on the fifty most important botanicals identified by the

European Union as being the most widely used in European countries, plus an additional thirty-five species identified by IBIDS staff. This list includes all the best-selling herbs, such as St. John's wort, echinacea, valerian, and ginkgo.

One useful feature of IBIDS is the ability to restrict a literature search to peer-reviewed journals (see Figure 13.2), which includes the American Botanical Society's *HerbalGram.* At present, the database contains journal articles from over 2,900 journals but intends to increase its coverage in the future. The site also features an online list of all journals in the database.

Medicinal Plants of Native America Database (MPNADB) <http://ars-genome.cornell.edu/Botany/>

Developed by Dr. Daniel E. Moerman, Professor of Anthropology at the University of Michigan, Dearborn, this database is based on a two-volume book of the same name containing 17,634 items representing the medicinal uses of 2,147 species from 760 genera and 142 families by 123 different Native American groups. This has basically the same search features as EthnobotDB. Data can be retrieved using the botanical name, common name, and action/uses (e.g., cold remedy, burn dressing, heart medicine). The URL goes to the Botanical Database section of the USDA-ARS Center for Bioinformation and

FIGURE 13.2. Representative Record from the IBIDS Database

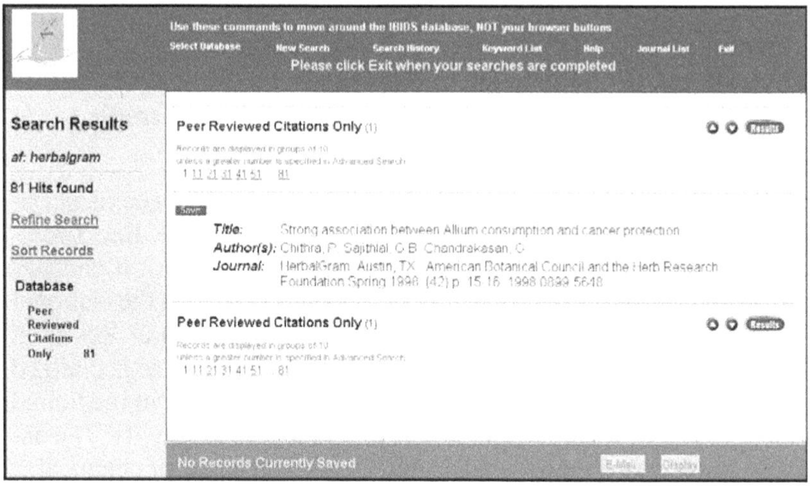

Comparative Genomics. Click on the MPNA DB link to enter the database.

MEDLINE (See PubMed)

Native Health History Database
<http://hsc.unm.edu/nhhd/>

The Native Health History Database (NHHD) is a computerized information resource containing complete bibliographic information and abstracts on historical American Indian and Alaska Native (AI/AN) medical/health research reports. The database is produced by the Indian Health Service (IHS) and the University of New Mexico (UNM) Health Sciences Center (HSC). Database entries cover the period from 1652 to 1970. The database contains citations on traditional foods and herbal remedies and is a useful resource for locating information on the use of medicinal plants among American Indian communities.

ODS Databases: Computer Access to Research on Dietary Supplements (CARDS)
<http://ods.od.nih.gov/databases/cards.html>

Produced by the Office of Dietary Supplements (ODS), this is a database of existing and ongoing dietary supplements research currently supported by federal agencies.

PhytochemDB/Phytochemical Database
<http://genome.cornell.edu/Botany/aboutphytochemdb.
html>

PhotochemDB is an abridged form of the original Phytochemical Database developed by Dr. James A. Duke and Stephen M. Beckstrom-Sternberg of the U.S. Department of Agriculture (USDA); it contains only the phytochemical and taxonomic data. It is housed at the National Germplasm Resources Laboratory (NGRL), Agricultural Research Service (ARS), U.S. Department of Agriculture. This is a specialized database devoted to providing data on chemicals found in medicinal plants, including quantity, taxonomic occurrence, and activity. The database can be searched by genus, family, common name, or chemical.

Plant Abstracts (Southwest School of Botanical Medicine)
<http://chili.rt66.com/hrbmoore/Abstracts/Abstracts.html>

This collection of citations and abstracts on European and Asian plant research is provided by the Southwest School of Botanical Medicine (see the corresponding entry in Chapter 3). Abstracts can be browsed by genus name (the first part of the botanical name).

POPLINE
<http://igm.nlm.nih.gov/>

In many traditional healing systems, herbal medicines are commonly used during pregnancy and childbirth. POPLINE is a bibliographic database maintained by the Population Information Program at the Johns Hopkins University with assistance from the Population Index, Princeton University. It provides worldwide coverage of population and family planning activities, women's health issues, as well as related developing country issues, such as AIDS and other sexually transmitted diseases, primary health care, population, and environment. As of this writing, this database is accessible through the NLM's Internet Grateful Med (IGM) system. Future plans include the incorporation of journal records into the PubMed system (see PubMed entry in Chapter 13) and the eventual replacement of IGM by a new NLM Gateway <http://gateway. nlm.nih.gov/gw/Cmd>.

PubMed (MEDLINE)
<http://www.ncbi.nlm.nih.gov/PubMed/>

The MEDLINE database, produced by the U.S. National Library of Medicine (NLM), is the primary source for current information from the biomedical literature, both in North America and in European countries. MEDLINE is the electronic version of the *Index medicus*. It covers the fields of medicine, nursing, dentistry, veterinary medicine, the health care system, and the preclinical sciences, such as anatomy and biochemistry. It currently contains references to around 11 million articles, from over 4,000 journals, dating back to 1966. Around 33,000 new citations are added each month! However, not every journal is indexed from cover to cover, and some are "selectively indexed"—if an article is not related to biomedicine it is not added to MEDLINE.

Since June 1997, MEDLINE has been available free on the Internet through the PubMed site. The NLM, as a government agency, was given a congressional mandate to make access to MEDLINE free. The PubMed system began as the bibliographic component of Entrez, a collection of databases from the NLM's National Center for Biotechnology Information (NCBI), designed for the molecular biology community. It is important to understand the distinction between PubMed and MEDLINE. PubMed includes citations from Pre-MEDLINE, an "in-process" database that provides basic citation information and abstracts before the citation is indexed by NLM and added to MEDLINE. In addition, many publishers have made agreements with NLM to provide electronic citations and abstracts as soon as a journal issue is published. This basic information is then transferred directly into the PubMed database without any of the usual indexing processes. There is little selective indexing, so when citations from the multidisciplinary journal *Science* are added to PubMed, those in disciplines such as astronomy and geology are added as well.

The key to effective MEDLINE searching is an understanding of how the database is indexed. All records in the MEDLINE database consist of two parts: a bibliographic citation and a list of indexing terms (see Figure 13.3). The bibliographic citation contains everything you need to know to locate the original article, that is, title, author(s), and journal information. However, the citations themselves often do not contain enough information to allow you to find them easily in a database as large as MEDLINE. To add more information to a citation, professional indexers at the NLM add indexing terms to the subject field of the MEDLINE record. The indexers do not randomly select terms; instead they choose from an approved list called a controlled vocabulary. Indexers decide which terms best describe the article's contents as well as the issues, topics, or concepts it conveys. Each article is then assigned up to fifteen indexing terms. These standardized medical terms are called medical subject headings (MeSH). The MeSH vocabulary is a controlled thesaurus of almost 17,000 terms maintained by the National Library of Medicine (NLM). As important new concepts appear in the literature, or as significant modifications of existing concepts appear, NLM assigns new MeSH headings.

Currently, MEDLINE lists around twenty-five main headings under Alternative Medicine, but many alternative medicine terms have

FIGURE 13.3. PubMed (MEDLINE) Record

: *Clin Otolaryngol* 1999 Jun;24(3):164-7 Related Articles, Books, LinkOut

⑤ online

Ginkgo biloba for tinnitus: a review.

Ernst E, Stevinson C

Department of Complementary Medicine, School of Postgraduate Medicine and Health Sciences, University of Exeter, UK. E.Ernst@exeter.ac.uk

Publication Types:
- Clinical trial
- Randomized controlled trial
- Review
- Review, tutorial

MeSH Terms:
- Antioxidants/therapeutic use*
- Dose-Response Relationship, Drug
- Ginkgo biloba/therapeutic use*
- Human
- Severity of Illness Index
- Tinnitus/drug therapy*
- Tinnitus/diagnosis

Substances:
- Antioxidants

yet to be included. Many herbal remedies are indexed in MEDLINE using the rather nebulous MeSH term "Plant Extracts," rather than any specific CAM term. The National Library of Medicine (NLM), however, in conjunction with the National Center for Complementary and Alternative Medicine (NCCAM), is committed to improving the indexing of CAM articles: in 2000, new headings were added for several popular herbs, including echinacea and feverfew. Table 13.1 lists many of the current MeSH terms that will retrieve articles related to herbal medicine. Effective searching, though, should employ a combination of MeSH terms and keywords.

PubSCIENCE
<http://pubsci.osti.gov/>

PubSCIENCE, produced by the Office of Scientific and Technical Information (OSTI) in the Department of Energy (DOE), is designed to facilitate access to peer-reviewed journal literature in the physical

TABLE 13.1. Medical Subject Headings for Herbs in MEDLINE

General Topics	Specific Herbs
Materia Medica	Aloe
Medicine, Ayurvedic	Areca
Medicine, Chinese Traditional	Capsicum
Medicine, Traditional	Chamomile
Ethnobotany	Cat's Claw
Ethnopharmacology	Comfrey
Herbs	Echinacea
Medicine, Herbal	Feverfew
Plants, Medicinal	Garlic
Plants	Ginkgo biloba
Drugs, Chinese Herbal	Ginseng
Plant Extracts	Hypericum
	Kava
	Milk Thistle
	Rhamnus
	Rhubarb
	Valerian

sciences and other energy-related disciplines. It includes many relevant research articles, with information on the drying and processing of herbal plant material, on the purification of plant compounds, and on plants that protect against the harmful effects of radiation, as well as biophysical studies on the molecular constituents of herbal plants.

The Scientific World—SciBase
<http://www.scientificworld.com/>

The Scientific World's SciBase database is a new Internet publishing venture providing online resources for researchers, including chat rooms and a peer-reviewed publishing component. It is a free searchable database of around 20,000 journals, including over 7,000 in the life sciences. SciBase claims that approximately 2,000 new articles are added daily. Links are provided to the article's abstract, where available, and copies of the complete article may be purchased online. Searching on "herb" retrieved over 100 citations — many in European and Asian journals not covered by MEDLINE, such as the *Chinese Journal of Sports Medicine*, the *Journal of Essential Oil Research,* and the *Korean Journal of Food Science and Nutrition.*

TOXNET (Toxicology Data Network)—TOXLINE
<http://toxnet.nlm.nih.gov/>

TOXLINE is a National Library of Medicine (NLM) bibliographic database that contains references to literature on the pharmacological, biochemical, physiological, and toxicological effects of drugs, environmental pollutants, and other chemicals. TOXLINE is now available, free of charge, on the NLM's Toxicology Data Network (TOXNET).

UIC/NIH Center for Botanical Dietary Supplements Research
<http://www.uic.edu/pharmacy/research/diet/>

NAPRALERT

This online resource requires a subscription, but it is included because it is a well-known database with in-depth information on many herbs, including Chinese herbs. NAPRALERT (NAtural PRoducts ALERT) is a large relational database started in 1975 by Dr. Norman Farnsworth and originally was designed to be used in drug discovery and development processes.[8] The database currently contains extracted information from over 150,000 scientific research articles, dating back to 1650. It includes both bibliographic and factual data and is one of the most comprehensive sources of information on natural products, including information on the pharmacology, biological activity, taxonomic distribution, ethnobotanical uses, and chemistry of plant extracts. In addition, considerable amounts of data on the chemistry and pharmacology (including clinical trials) of naturally occurring secondary metabolites are stored in the database. NAPRALERT is maintained by the Program for Collaborative Research in the Pharmacological Sciences (PCRPS), College of Pharmacy, University of Illinois at Chicago. PCRPS is a World Health Organization Collaborating Center for Traditional Medicines. For information about the database, choose the NAPRALERT link from the home page.

United States Patent and Trademark Office (USPTO)
Web Patent Databases
<http://www.uspto.gov/patft/>

See the entry for European Patent Office (EPO) in this chapter.

University of Maryland Complementary Medicine Program
<http://www.compmed.ummc.umaryland.edu/>

Complementary and Alternative Medicine Pain Database (CAMPAIN)

CAMPAIN, a bibliographic database created by the University of Maryland's Complementary Medicine Program (CMP), is designed to collect and compile the results of scientific research relating to the use of CAM in the treatment of pain and pain-related conditions (see the University of Maryland CMP entry in Chapter 6 for more information). Access the database by selecting Bibliographic Databases from the home page. The database contains citations for approximately 10,000 research reports from around the world concerning pain and complementary medicine.

Chapter 14

AIDS, Cancer, and Other Specialized Health Areas

The entire economy of the Western world is built on things that cause cancer.

From the 1985 movie *Bliss*

COMPLEMENTARY OR ALTERNATIVE MEDICINE

In discussing the use of CAM therapies by patients with life-threatening conditions such as AIDS or cancer, it is useful first to clarify the distinction between complementary medicine and alternative medicine. The terms *complementary medicine* and *alternative medicine* are often used interchangeably, but there are important distinctions. Alternative therapies are just that, and these are usually promoted as being valid treatment options apart from those offered by conventional medicine, indicating that either treatment approach might be effective. Complementary, however, suggests the use of therapies that can be used alongside mainstream medical treatments. Ginger *(Zingiber officinale),* for example, is often advocated as being beneficial for reducing nausea and vomiting. It is currently being investigated for the relief of postoperative and chemotherapy-induced nausea.[1,2]

CANCER AND CAM

It is estimated that approximately 8 to 10 percent of cancer patients seek some type of alternative treatment.[3] Two of the best-known, and controversial ones, are the Hoxsey and Essiac herbal treatments for cancer. Essiac is named after a Canadian nurse, Rene M. Caisse ("Essiac" is Caisse spelled backwards!), and is probably the most popular of all alternative treatments for cancer. It is essentially an herbal tea, reputedly based on an American Indian formula. It originally contained four herbs: burdock root *(Arctium lappa),* sheep sorrel *(Rumex acetosella),* slippery elm bark *(Ulmus fulva),* and turkey rhubarb *(Rheum palmatum)* or Indian rhubarb *(Rheum officinale).* The Hoxsey treatment is named after Harry Hoxsey (1901-1974), a self-taught healer who claimed to have cured many cancer patients using an herbal remedy handed down by his great-grandfather. Up until the 1950s, Hoxsey clinics could be found in several U.S. states, but the only surviving one is now in Tijuana, Mexico. (For details of the controversies surrounding these two cancer treatments, see the report *Unconventional Cancer Treatments* discussed later in this chapter.)

CAM AND HIV/AIDS

Many people with AIDS/HIV use CAM therapies. A recent study reported that around 56 percent of people with AIDS/HIV in Australia used some type of CAM therapy, often as a complementary treatment to alleviate side effects accompanying the use of antiretroviral drugs.[4] Many studies are currently underway or being developed to assess the efficacy of CAM treatments in people with HIV/AIDS. In September 1994, the National Center for Complementary and Alternative Medicine (NCCAM) awarded $840,000 to Bastyr University in Seattle to study CAM treatments for HIV/AIDS (see the university's entry in this chapter). Herbs that have been used in the management of HIV disease include tea tree oil (for fungal infections), garlic (an extract called allicin for cryptosporidiosis), sage (for night sweats), slippery elm (for diarrhea), and echinacea (for HIV infection).

As the popularity of CAM continues to grow, many studies are being initiated to test its efficacy in the treatment of a variety of diseases and medical or health conditions. In addition to cancer and AIDS/HIV, this chapter brings together Web sites for other important diseases and health areas, such as diabetes, Alzheimer's disease, and women's health.

WEB SITES THAT FOCUS ON PARTICULAR DISEASES AND HEALTH ISSUES

The Alzheimer's Association
<http://www.alz.org/>

The Alzheimer's Association is the largest national voluntary health organization committed to finding a cure for Alzheimer's disease and to helping those already affected. The association also provides education and support for people diagnosed with the condition, their families, and caregivers. Several herbs, such as ginkgo biloba, are being widely studied for their effect on Alzheimer patients.

The American Dietetic Association (ADA)
<http://www.eatright.org/>

The ADA is the world's largest organization of food and nutrition professionals, with nearly 70,000 members in fifty-seven countries. ADA members serve the public by offering prevention and wellness services and medical nutrition therapy in a variety of settings, including health care, food service, business and communications, research, education, and private practice.

A new category of foods, functional foods, has recently appeared. This category includes a large number of herbal-enriched products that make a variety of structure/function claims. Examples include cereal fortified with ginkgo biloba, which is marketed as reducing symptoms of dementia and juice fortified with echinacea, marketed for boosting the immune system.

Bastyr University AIDS Research Center (BUARC)
<http://www.bastyr.edu/research/buarc/>

BUARC was established in 1994 under a cooperative grant from NIH's National Institute of Allergy and Infectious Disease (NIAID) and the National Center for Complementary and Alternative Medicine (NCCAM). Its mission is not only to screen and evaluate CAM therapies but to provide consultation and support to the medical and research community in the scientific evaluation of CAM therapies. A pilot grant program is currently funding studies on a number of botanical extracts, including Chinese herbs and licorice root *(Glycyrrhiza glabra).*

Bulletin of Experimental Treatments for AIDS (BETA)
<http://www.sfaf.org/beta/index.html>

BETA is published by the Treatment Education and Support Unit of the San Francisco AIDS Foundation. It covers new developments in AIDS treatment research, including in-depth articles on treatment for HIV infection and AIDS-related illnesses for HIV-positive individuals and their caregivers. To access relevant articles, select Browse by Topic, then Alternative and Complementary Treatments, or type a keyword into the search box.

The Cancer Information Service (CIS)
<http://cis.nci.nih.gov/about/about.html>

The Cancer Information Service (CIS) is part of the federal government's National Cancer Institute (NCI). It provides cancer information for patients, their families, the general public, and health professionals. In addition, CIS regional offices work with research partners in local cancer control and health communications research projects. From the home page, go to Cancer Resources to access documents on cancer and complementary and alternative medicine (CAM) therapies.

CancerNet
<http://cancernet.nci.nih.gov/>

CancerNet contains material for health professionals, patients, and the public, including information about cancer treatment, screening, prevention, supportive care, and clinical trials, as well as CANCER-LIT, a bibliographic database. Under Treatment Options, select Complementary and Alternative Medicine. There is a useful Questions and Answers section about the use of CAM in cancer treatment.

Center for Alternative Medicine Research in Cancer:
University of Texas
<http://www.sph.uth.tmc.edu/utcam/default.htm>

The University of Texas Center for Alternative Medicine Research specializes in the study of alternative and complementary therapies for cancer prevention and control. (For more details, see the corresponding entry in Chapter 6.)

Cyberbotanica: Plants and Cancer Treatment
<http://biotech.icmb.utexas.edu/botany/>

This resource was developed by Lucy Snyder as part of the Indiana University Biotech Project. It provides information on the various botanical compounds currently used in cancer treatment and the plants that produce them.

The Electronic Textbook of Dermatology—Botanical Dermatology
<http://telemedicine.org/stamford.htm>

This is a chapter from the *The Electronic Textbook of Dermatology,* a project of the Internet Dermatology Society (IDS). The IDS is dedicated to facilitating access to dermatology information resources and to promoting uniform access to dermatology educational materials. Select Botanical Dermatology from the table of contents.

HIV InSite
<http://hivinsite.ucsf.edu/>

HIV InSite is a project of the University of California San Francisco (UCSF) Positive Health Program at San Francisco General Hospital Medical Center and the UCSF Center for AIDS Prevention Studies. It is one of the most comprehensive collections of resources for HIV/AIDS on the Internet.

JAMA HIV/AIDS Resource Center
<http://www.ama-assn.org/special/hiv/>

Produced by the *Journal of the American Medical Society* (JAMA), this is one of the most respected Internet sites of information on HIV/AIDS, with a large collection of peer-reviewed publications and resources.

Menopause Online
<http://www.menopause-online.com/>

Menopause Online's mission is to provide women with up-to-date, easy-to-use information managing symptoms associated with menopause. This site covers several herbal therapies that show promise in relieving some of the common discomforts. For example, black cohosh is approved in Germany to treat hot flashes.[5] This site has a useful list of menopause symptoms, with information on specific herbs that may help relieve a particular set of symptoms.

National Cancer Institute (NCI)
<http://www.nci.nih.gov/>

The National Cancer Institute (NCI), a component of the National Institutes of Health (NIH), is the federal government's principle agency for cancer research and training. Established under the National Cancer Act of 1937, NCI has a mandate to conduct and support research, training, health information dissemination, and other programs relevant to the cause, diagnosis, prevention, and treatment of cancer. NCI services include CancerNet and the Cancer Information Service (see the corresponding entries in this chapter).

The National Women's Health Information Center
<http://www.4woman.gov/>

The National Women's Health Information Center is part of the Office on Women's Health (OWH), within the Department of Health and Human Services (DHHS). OWH's mission is the development and implementation of new programs and initiatives to improve women's health not only in the United States but internationally. The center provides links to sites with herbal information relevant to women's health issues. For example, black cohosh tea has been used in Native American cultures for centuries to alleviate the symptoms associated with menopause.

OncoLink
<http://cancer.med.upenn.edu/>

OncoLink Complementary Medicine
<http://www.oncolink.upenn.edu/specialty/complementary/>

The University of Pennsylvania's OncoLink is one of the most respected medical sites for cancer information on the Internet. It provides excellent information for both the health professional and the consumer. A special Complementary Medicine section provides comprehensive information on all aspects of CAM and cancer. Materials include books, videos, and audiotapes.

Prostate Cancer Program (University of Michigan Comprehensive Cancer Center)
<http://www.cancer.med.umich.edu/prostcan/prostcan.html>

This site from the University of Michigan is a resource for information pertaining to prostate cancer, such as diagnosis, staging, treatment options, available specialists, investigational studies, and current research. It is intended for both patients and health professionals. Information on herbal therapies such as saw palmetto is included.

SusanLoveMD.com: The Website for Women
<http://www.susanlovemd.com/>

Susan Love, MD, is an author, teacher, surgeon, researcher, and activist for women's health issues. Go to the section Complementary and Alternative Therapies/Herbal Remedies and Herbs for Breast Cancer.

Unconventional Cancer Treatments (Office of Technology Assessment [OTA])
<http://www.wws.princeton.edu/~ota/>

The U.S. Office of Technology Assessment (OTA) closed on September 29, 1995. During its twenty-three-year history, OTA provided U.S. congressional members and committees with objective and authoritative analysis of the complex scientific and technical issues of the late twentieth century. It was a leader in practicing and encouraging delivery of public services in innovative and inexpensive ways. *Unconventional Cancer Treatments* contains useful information on herbal treatments and is still sometimes quoted in the literature. Though this report is from 1990, it contains a very balanced account of herbal cancer treatments, including the controversial Hoxsey, Essiac, and mistletoe herbal treatments

University of Medicine and Dentistry of New Jersey (UMDNJ) School of Health Related Professions (SHRP)
<http://www.umdnj.edu/shrpweb/nutr/Oralherb%20index.html>

Several herbs are used by herbalists as part of oral hygiene or to treat gum diseases. This site has information on such herbs as rhatany *(Krameria triandra),* northern prickly ash, or "the toothbrush tree"

(Zanthoxylum americanum), and bloodroot *(Sanguinaria cana-densis).* The latter herb is widely used in the treatment of gingivitis.

University of Pittsburgh Voice Center—Taking Herbal Medicines: What Singers Should Know <http://www.upmc.edu/UPMCVoice/Herbmedsingers.htm>

The University of Pittsburgh Voice Center is a unique resource for both the general public and professional singers. Since some CAM therapies are being used to treat voice defects and may be harmful, this site provides information for singers regarding the use of herbal medicines, vitamins, and nutrients.

Chapter 15

Aromatherapy

The rose looks fair, but fairer we it deem
For that sweet odour which doth in it live.

Shakespeare, Sonnet LIV

WHAT IS AROMATHERAPY?

Aromatherapy is a division of herbal medicine that uses fragrant oils extracted from the roots, branches, bark, leaves, fruit, and flowers of various plants to enhance and improve health. It has its roots in ancient medical practice dating back to Egyptian times, although it was not called aromatherapy until 1928, when the term was coined by a French chemist named Reneé Gattefossé, who documented the healing benefits of essential oils and wrote the first modern book on the topic.[1] The word *aromatherapy* is a little misleading, since treatment is not just limited to inhaling vapors; oils are often used during a therapeutic massage or simply added to bathwater.

Each of the 130 or so varieties of essential oils used by the aromatherapist is believed to have specific healing properties that can help in the treatment of many ailments, ranging from relief from stress and minor physical conditions, such as sore throats and colds, to more serious ones, such as bronchitis. Some claim that these oils can relieve bacterial infections, immune disorders, cystitis, and herpes simplex, and that they even help patients with life-threatening conditions such as cancer and AIDS.[1] Examples of commonly used oils are lavender *(Lavendula officinalis),* having antidepressant, anti-inflammatory, and antibacterial effects; eucalyptus *(Eucalyptus globules),* acting as a stimulant and decongestant; and rosemary *(Rosmarinus officinalis),* to stimulate the circulatory system and soothe aching muscles.[1]

AROMATHERAPY IN EUROPE

Aromatherapy is growing in popularity in the United States but is already fairly well established in England, France, Switzerland, and New Zealand, with all four countries having professional licensing standards. (In the United States, there is currently no government-recognized certification program in aromatherapy and essential oils are not regulated by the FDA or any other agency.) In Europe, aromatherapy is widely used by nurses and other care providers to improve mood and behavior in cancer, HIV, and Alzheimer patients. Some evidence suggests that peppermint and cardamom oil relieve the nausea brought on by chemotherapy.[2] Current research in England is studying the use of oil from lemon balm *(Melissa officinalis)* to improve failing memory and to treat dementia in patients with neurological conditions such as Alzheimer's disease.[3] Aromatherapy is reported to have a calming effect on these patients, and evidence suggests that components of balm oil interact with chemicals (the neurotransmitters) in the brain.[3]

It should be emphasized that aromatherapy does not include the Bach flower remedies, a complex system of flower healing developed in the 1930s by British physician Edward Bach (1886-1936). The Bach remedies involve flower extracts and their "essence" or "life force" to improve the mental and emotional well-being of a patient.[4] The Bach flower healing system is more closely related to homeopathy than herbalism, since it uses the principle of extremely minute forms of medication to prescribe the most appropriate remedy.[4]

WEB SITES PROVIDING INFORMATION ON AROMATHERAPY

AGORA: Aromatherapy Global Online Research Archives <http://www.nature-helps.com/agora/agora.html>

The AGORA Web site is an international group of volunteers dedicated to providing noncommercial aromatherapy education articles via donated Web space on the Internet. The site is maintained by Michel Vanhove.

Aromatherapy Trade Council
<http://www.a-t-c.cwc.net/>

The Aromatherapy Trade Council is an organization based in the United Kingdom. Its mission is to promote responsible marketing and consumer safety.

AromaWeb
<http://www.aromaweb.com/>

This is a comprehensive collection of resources with basic information on the science and practice of aromatherapy. Of particular note is the Oil Profile section, providing profiles of seventy commonly used essential oils. The site also provides safety information about aromatherapy practices and products.

The Guide to Aromatherapy
<http://www.fragrant.demon.co.uk/>

This is a popular site for basic information on aromatherapy, maintained by Graham Sorenson from Wales. Of special interest are his descriptions of the individual oils and some safety information for specific oils.

International Federation of Aromatherapists (IFA)
<http://www.int-fed-aromatherapy.co.uk/>

The International Federation of Aromatherapists (IFA) is based in the United Kingdom and is the oldest international aromatherapy organization. Since its foundation, it has pioneered the use of aromatherapy in hospitals, hospices, special care units, and general practice. IFA maintains a searchable database of aromatherapy practitioners.

Medicinal Plant Research Centre (United Kingdom)
<http://www.newcastle.ac.uk/medplant/>

In collaboration with the Mental Health Foundation of the United Kingdom, the center is currently conducting clinical trials involving the use of oil from lemon balm *(Melissa officinalis)* to treat patients with dementia. (For more information about the center, see the corresponding entry in Chapter 6.)

The Mental Health Foundation
<http://www.mentalhealth.org.uk/>

Bibliography (The Use of Aromatherapy to Treat Mental Health Problems)
<http://www.mentalhealth.org.uk>

The Mental Health Foundation, based in the United Kingdom, is a nonprofit charity group working to improve services for people with mental health problems or learning disabilities. It funds and works with both service users and health care providers and plays an important role in funding research and investigating new approaches to prevention, treatment, and care in the United Kingdom. The foundation is investigating CAM therapies to help control depression and dementia, including aromatherapy, and is sponsoring clinical trails to investigate the effect of essential oils on patient mood and behavior (see also the previous entry for the Medicinal Plant Research Centre).

The foundation publishes an online Catalogue of Research Materials, including a useful annotated bibliography of publications on the use of aromatherapy in health care. To view, type the terms Research Materials into the search engine box available at the home page, then select catalog of research materials, then aromatherapy.

The National Association for Holistic Aromatherapy (NAHA)
<http://www.naha.org/>

The NAHA is a nonprofit U.S. organization dedicated to enhancing public awareness of the benefits of aromatherapy, the medicinal use of aromatic plants and essential oils. Its mission includes promoting and elevating academic standards in aromatherapy education and practice.

Chapter 16

Mailing Lists, Newsgroups, and Chat Rooms

If you have knowledge, let others light their candles at it.

Margaret Fuller

THE INTERNET AND THE NEW COMMUNICATION CHANNELS

The Internet not only is providing ready access to a wealth of health-related information but is beginning to transform radically the way health care is delivered.[1] Ubiquitous access to vast quantities of online medical information is already producing a more knowledgeable group of consumers. This will most likely create a more equitable relationship between the educated consumer and the health professional, as patients routinely use the Internet to check on what the doctor is telling them. The Internet is also effecting other changes, for one of its most innovative features is the number of new forms of two-way communication that it provides, with new opportunities for users to exchange information. For many people, the interactivity of these new channels of communication is one of the Internet's chief attractions. Electronic mail, newsgroups, and chat rooms have opened up new avenues for consumers to interact not only with their physicians but with one another.

Three of the most important electronic communication channels are mailing lists, newsgroups, and online chat rooms. Mailing lists, often erroneously referred to as "listservs," after one of the software programs that can be used to run them, require only an Internet connection and an electronic (e)-mail client program. An interested person simply subscribes to a listserv by sending an e-mail message to a special address. The person is then added to the mailing list and receives messages that are sent to the entire group. A list can be moderated or unmoderated. In a moderated list, an "owner" (operator) reads all incoming messages before they are posted to the group. When you subscribe, your name and e-mail address are added to the list and you usually receive a welcoming e-mail and begin receiving all postings sent by other members. You generally have the option of receiving messages individually or as a digest.

Newsgroups are essentially online bulletin boards, and these are also referred to as "Usenet." There are thousands of individual newsgroups, variously called conferences, forums, bulletin boards (BBs), or special interest groups (SIGs). They differ from mailing lists primarily in that messages are not automatically mailed to a person's e-mail account, but instead the user must access the bulletin board in much the same way as he or she would access a Web page. Each user logs on to the news server and views a list of messages, which can be downloaded and read, and an individual can post a message to be read by other members of the newsgroup. An extremely useful feature of bulletin boards is the "thread," whereby messages are grouped together around an original posting that raises a question or defines a new subject for discussion.

There are two main ways to access newsgroups. You can either access them using the World Wide Web, via your browser, or you can employ a "News Client" program, also known as a "newsgroup reader." The DejaNews site is the best known Web-based news reader and is essentially a database of over 500 million archived messages dating back to 1995. You can search the DejaNews database for particular articles or review threaded articles on a particular topic. Alternatively, newsgroups can be accessed using what is known as "reader software," or a "News Client." If you are using client software, you will be supplied with access to Usenet newsgroups by your Internet Service Provider (ISP). If your browser is part of a collection of bundled internet software it will already have a "News Client:" for Netscape Communicator the client is "Netscape News," while Microsoft Internet Explorer has "Internet News." The most useful active newsgroup for herbal information is alt.folklore.herbs.

Mailing lists and bulletin boards are often described as being "asynchronous," in that users who post messages and those who receive them do not have to be communicating at the same time. A user may post and receive messages at any time during the day or night. The third type of important Internet communication tool, the chat room, is different; communication takes place in real time, requiring participants to be online at the same time. Internet chat rooms are now widely used by millions of people. "Internet chat" used to be synonymous with Internet Relay Chat (IRC), a system whereby participants use special IRC software to talk with one another and log onto IRC channels. Other types of chat room systems have appeared in recent years though, and they are increasingly being integrated into the World Wide Web, so that a regular Web browser is often the only software needed.

INTERNET SITES FOR MAILING LISTS, NEWSGROUPS, AND CHAT ROOMS

Algy's Herb Page
<http://www.algy.com/herb/>

Algy's Herb Page is essentially an electronic bulletin board for information on the use of herbs (see entry in Chapter 4).

Allergy Discussion Group
<http://www.immune.com/allergy/>

The allergy mailing list is an unmoderated, consumer-oriented mailing list for the discussion of all types of human allergies, including treatments for allergies, allergy self-care, and support systems. Go to the given URL for subscription details. This site also provides an archive of messages that can be searched by keyword.

Alternative Medicine Research Mailing List

This list was created to promote discussion and collaboration in the emerging field of alternative medicine research. Pertinent topics include new findings, methodological issues, announcements of upcoming research conferences and seminars, and requests for information or collaboration regarding proposed studies. To subscribe to the list, send an e-mail message to <majordomo@Virginia.edu>. In the body of the message, type "subscribe altmed-res." The mailing list owner is <mbm2y@_eadings.edu> (Martha Brown Menard).

Alternatives for Healthy Living
<http://www.alt-med-ed.com/>

This Web site is designed to be an interactive community, featuring searchable forums, message boards, chat rooms, and reference libraries for most areas of complementary and alternative medicine.

Anti-Quackery Mailing Lists: "Healthfraud" Discussion List
<http://www.quackwatch.com/00AboutQuackwatch/discuss. html>

The Healthfraud discussion list is sponsored by the Georgia Council Against Health Fraud, Inc., and Quackwatch, Inc., both affiliates of the National Council Against Health Fraud (NCAHF). This list is devoted to the discussion of "unscientific" medical practices and quackery. The given URL provides access to information on how to join and use the list.

Aromatherapy List

To join the Aromatherapy List, send an e-mail to <listserv@idma. com>. Type "subscribe aromatherapy" in the body of the message.

Deja.com
<http://www.deja.com>

Google Groups
<http://groups.google.com/>

Formerly called DéjaNews, this searchable archive of newsgroup messages is a valuable resource for identifying pertinent newsgroups. You can also browse by topic or search messages by keywords. As of this writing, Deja.com's Usenet Discussion Service has been acquired by Google Inc., and a new interface is in development.

Henriette's Herbal Homepage
<http://www.ibiblio.org/herbmed/>

This popular site began life as a list of frequently asked questions for the Usenet newsgroup <alt.folklore.herbs>. Three lists are featured at this site: The Medicinal Herblist, The Culinary Herblist, and The HerbInfo-list archives. The site also features a newsgroup: <alt.folklore.herbs>.

Herbal Hall
<http://www.herb.com/herbal.htm>

This interactive Web site for professional herbalists features a Web-based discussion group primarily focused on "adaptogenic, nutritive, food or medicinal herbs." Click on HerbMail to go the relevant section.

HerbInfo
<http://alist4u.net/herbinfo.html/>

HerbInfo is an unmoderated mailing list, intended to be an all-purpose general discussion list for herbs. To subscribe, send e-mail to <herbinfo-request@bolis.com>. In the body of the message, type "subscribe." Archives of messages are available at the given URL.

The Herblist: The Medicinal and Aromatic Plants Discussion List
<http://www.ibiblio.org/herbmed/archives/herblist/rules.html#this-list/>

The Herbalist deals with issues associated with cross-cultural medicine and/or folk/herbal medicine. It used to be supported by staff at the Anadolu University Medicinal Plants Research Center in Turkey, but has now moved to a different server. The new list owner is Henrietta Kress (see the entry for Henrietta's Herbal Homepage).

Holistic-L
<http://www.scils.rutgers.edu/~mavcol/holist-l.html>

This is an unmoderated list for such health professionals as medical doctors, researchers, nurses, and physical therapists to discuss alternative health concepts and patient care. To subscribe to the list, send an e-mail message to <Listserv@Citadel.Net>. In the body of the message, type "Join Holistic-L."

Liszt, the Mailing List Directory
<http://www.liszt.com/>

Liszt is a user-friendly directory for browsing or searching for mailing lists that discuss certain subjects. Thousands of mailing lists exist, and though only a few are devoted specifically to the discussion of herbal topics, useful information can often be found in some of the more general health-related lists, especially those devoted to specific medical conditions or diseases. Search by keywords or browse by subject headings.

Natural Healthline
<http://www.naturalhealthvillage.com/>

This is an online newsletter of events and information from Natural HealthLine and the Natural Health Village. To subscribe, send an e-mail message to <webmaster@naturalhealthvillagevillage.com>. In the body of the message, type "subscribe healthline (your name)." For more information, go to the given URL.

Office of Dietary Supplements (ODS) Listserv
<http://ods.od.nih.gov/maillist.html>

The ODS supports research and disseminates research results in the area of dietary supplements (see the ODS entry in Chapter 8). The ODS listserv is a convenient way to receive news from the ODS through e-mail containing information about the ODS Web site, information on the availability of new fact sheets on dietary supplements, and other news, such as updates to the IBIDS database (see the IBIDS entry in Chapter 13).

Paracelsus — Clinical Practice in the Healing Arts
<http://www.healthwwweb.com/paracelsus.html/>

The Paracelsus discussion group is a service of the Integrative Medical Arts Group, Inc. The primary focus of this list is the clinical practice of integrative medicine, alternative therapies, and "natural" health care. Subscription is limited to health care practitioners, educators, researchers, and students in alternative and conventional medical fields and requires a short biographical note.

Phytopharmacognosy Internet Discussion Group
<http://www.phytochemistry.freeserve.co.uk/frames/ phyto.htm>

This is a moderated discussion group of experts in medicinal and aromatic plants. The list owner is Dr. John Wilkinson, Senior Lecturer in Phamacognosy, Botany, Organic and Phytochemistry at Middlesex University, United Kingdom. Membership is free and open to academics, industrialists, and other suitably qualified professionals who are involved in plant-based natural products and who have an e-mail address. Active membership is limited to experts only, and potential members must submit evidence of their expertise before they can participate. Discussions are focused on the botany and chemistry of medicinal and economic plants, ethnobotany, secondary metabolites, traditional medicines, volatile oils, tropical agriculture, academic and industrial problems, herbalism, aromatherapy, ecological biochemistry and evolution, and plant chemicals. It also includes notices on conferences and specialist topic discussions. Go to the above URL for further information.

Tibetan Plateau Project (TPP) Discussion List

The TPP sponsors an e-mail discussion list called "tpp-tibmed-plants" on the topic of medicinal plant conservation and the practice of Tibetan medicine. You can request subscription information by sending an e-mail message to tppei@earthisland.org with the phrase "subscribe tpp-tibmed-plants."

Notes

Preface

1. K. Parfitt. *Martindale: The Complete Drug Reference.* Thirty-Second Edition. London, UK: Taunton, MA: Pharmaceutical Press, 1999.

2. G. B. Wood. *The Dispensatory of the United States of America.* Twenty-Second Edition, with supplement. Philadelphia: Lippincott, 1940.

3. V. E. Tyler. *The Honest Herbal: A Sensible Guide to the Use of Herbs and Related Remedies.* Third Edition. Binghamton, NY: Pharmaceutical Products Press, 1993.

4. _____. *Herbs of Choice: The Therapeutic Use of Phytomedicinals.* Binghamton, NY: Pharmaceutical Products Press, 1994.

5. M. Blumenthal, W. R. Busse, and German Federal Institute for Drugs and Medical Devices. *The Complete German Commission E Monographs.* Austin, TX; Boston: American Botanical Council; Integrative Medicine Communications, 1998.

6. D. R. Hill. Brandon-Hill selected list of books and journals for the small medical library. *Bulletin of the Medical Library Association* 87(2): 145-169, 1999.

7. J. W. Spencer, and J. J. Jacobs. *Complementary Alternative Medicine: An Evidence-Based Approach.* St. Louis: Mosby, 1999.

8. D. J. Owen. Herbal resources on the Internet. *Medical Reference Services Quarterly* 18(4): 39-56, 1999.

Chapter 1

1. V. E. Tyler. *The Honest Herbal: A Sensible Guide to the Use of Herbs and Related Remedies.* Third Edition. Binghamton, NY: Pharmaceutical Products Press, 1993, p. xi.

2. T. P. I. A. L. Report. "The Online Health Care Revolution: How the Web Helps Americans Take Better Care of Themselves." The Pew Research Center, 2000.

3. W. M. Silberg, G. D. Lundberg, and R. A. Musacchio. Assessing, controlling, and assuring the quality of medical information on the Internet: Caveant lector et viewor—Let the reader and viewer beware. *Journal of the American Medical Association* 277(15): 1244-1245, 1997.

4. D. Nowak and T. Zlatic. Herbal products and the Internet: A marriage of convenience. *Journal of the American Pharmaceutical Association* 39(2): 241-242, 1999.

5. P. Kim, T. R. Eng, M. J. Deering, and A. Maxfield. Published criteria for evaluating health related Web sites: Review. *British Medical Journal* (Clinical Research Edition) 318(7184): 647-649, 1999.

Chapter 2

1. D. E. Moerman. *Medicinal Plants of Native America, Research Reports in Ethnobotany; Contribution 2.* Ann Arbor: Museum of Anthropology, University of Michigan, 1986.

2. W. Boyle and F. J. Brinker. *Herb Doctors: Pioneers in Nineteenth-Century American Botanical Medicine and a History of the Eclectic Medical Institute of Cincinnati.* East Palestine, OH: Buckeye Naturopathic Press, 1988.

3. W. Boyle. *Official Herbs: Botanical Substances in the United States Pharmacopoeias, 1820-1990.* First Edition. East Palestine, OH: Buckeye Naturopathic Press, 1991.

4. M. G. Regan-Smith. Reform without change: Update. *Academic Medicine* 73(5): 505-507, 1998.

5. V. E. Tyler. *The Honest Herbal: A Sensible Guide to the Use of Herbs and Related Remedies.* Third Edition. Binghamton, NY: Pharmaceutical Products Press, 1993.

6. D. M. Eisenberg, R. B. Davis, S. L. Ettner, S. Appel, S. Wilkey, M. Van Rompay, and R. C. Kessler. Trends in alternative medicine use in the United States, 1990-1997: results of a follow-up national survey. *Journal of the American Medical Association* 280(18): 1569-1575, 1998.

7. D. M. Eisenberg, R. C. Kessler, C. Foster, F. E. Norlock, D. R. Calkins, and T. L. Delbanco. Unconventional medicine in the United States: Prevalence, costs, and patterns of use. *New England Journal of Medicine* 328(4): 246-252, 1993.

8. B. A. Johnston. One-third of nation's adults use herbal remedies. Market estimated at $3.24 billion. *HerbalGram* 40(Summer): 49, 1997.

9. W. B. Jonas. Researching alternative medicine. *Nature Medicine* 3(8): 824-827, 1997.

10. J. S. Williamson and C. M. Wyandt. The herbal generation: Trends, products, and pharmacy's role. *Drug Topics* 143(7): 69-78, 1999.

Chapter 3

1. Frederick C. Mish, ed. *Merriam-Webster's Collegiate Dictionary.* Tenth Edition. Springfield, MA: Merriam-Webster, 2001.

2. N.G. Bisset. *Herbal Drugs and Phytopharmaceuticals.* Stuttgart; Boca Raton: Medpharm: CRC Press, 1994.

3. B. Jackson and A. Reed. Catnip and the alteration of consciousness. *Journal of the American Medical Association* 207(7): 1349-1350, 1969.

4. J. T. Petersik, J. Poundstone, and J. W. Estes. Of cats, catnip, and Cannabis. *Journal of the American Medical Association* 208-360, 1969.

5. N. R. Slifman, W. R. Obermeyer, B. K. Aloi, S. M. Musser, W. A. Correll Jr., S. M. Cichowicz, J. M. Betz, and L. A. Love. Contamination of botanical dietary supplements by *Digitalis lanata. New England Journal of Medicine* 339(12): 806-811, 1998.

6. A. Y. Leung and S. Foster. *Encyclopedia of Common Natural Ingredients Used in Food, Drugs, and Cosmetics*. Second Edition. New York: Wiley, 1996.

7. M. Blumenthal, W. R. Busse, and German Federal Institute for Drugs and Medical Devices. *The Complete German Commission E Monographs*. Austin, TX; Boston: American Botanical Council; Integrative Medicine Communications, 1998.

8. C. B. Clarke. *Edible and Useful Plants of California, California Natural History Guides; 41*. Berkeley: University of California Press, 1977.

9. L. Watson and M. J. Dallwitz. The families of angiosperms: Automated descriptions, with interactive identification and information retrieval. *Aust. Syst. Bot* 4: 681-695, 1991.

10. D. E. Moerman. *Medicinal Plants of Native America, Research Reports in Ethnobotany; Contribution 2*. Ann Arbor: Museum of Anthropology, University of Michigan, 1986.

Chapter 4

1. D. Hoffmann. *The Herbal Handbook: A User's Guide to Medical Herbalism* (p 19). Rochester, VT: Healing Arts Press, 1988.

2. B. A. Johnston. One-third of nation's adults use herbal remedies. Market estimated at $3.24 billion. *HerbalGram* 40(Summer): 49, 1997.

3. T. Johnson. *CRC Ethnobotany Desk Reference*. Boca Raton, FL: CRC Press, 1999.

Chapter 5

1. G. Benzi and A. Ceci. Herbal medicines in European regulation. *Pharmacological Research* 35(5): 355-362, 1997.

2. M. Blumenthal, W. R. Busse, and German Federal Institute for Drugs and Medical Devices. *The Complete German Commission E Monographs*. Austin, TX; Boston: American Botanical Council; Integrative Medicine Communications, 1998.

3. British Herbal Medicine Association, Scientific Committee. *British Herbal Pharmacopoeia, 1983*. Consolidated. London: The Committee, 1983.

Chapter 6

1. E. Ernst. Complementary medicine: from quackery to science? *Journal of Laboratory and Clinical Medicine* 127(3): 244-245, 1996.

2. K. R. Pelletier, A. Marie, M. Krasner, and W. L. Haskell. Current trends in the integration and reimbursement of complementary and alternative medicine by managed care, insurance carriers, and hospital providers. *American Journal of Health Promotion* 12(2): 112-122, 1997.

3. M. S. Wetzel, D. M. Eisenberg, and T. J. Kaptchuk. Courses involving complementary and alternative medicine at US medical schools. *Journal of the American Medical Association* 280(9): 784-787, 1998.

4. A. Toufexis. Dr. Jacobs' alternative mission. A new NIH office will put unconventional medicine to the test. *Time* 141(9): 43-44, 1993.

5. J. Couzin. "Beefed-Up NIH Center Probes Unconventional Therapies." *Science,* December 18: 2175. 1998

6. D. M. Eisenberg, R. C. Kessler, C. Foster, F. E. Norlock, D. R. Calkins, and T. L. Delbanco. *New England Journal of Medicine* 328(4): 246-252, 1993.

7. S. Barrett. The public needs protection from so-called 'alternatives'. *Internist* 35(8): 10-11, 1994.

8. D. Hoffmann. *The Herbal Handbook: A User's Guide to Medical Herbalism.* Rochester, VT: Healing Arts Press, 1988.

Chapter 7

1. N. R. Farnsworth, O. Akerele, A. S. Bingel, D. D. Soejarto, and Z. Guo. Medicinal plants in therapy. *Bulletin of the World Health Organization* 63(6): 965-981, 1985.

2. M. A. Pathak and T. B. Fitzpatrick. The evolution of photochemotherapy with psoralens and UVA (PUVA): 2000 BC to 1992 AD. *Journal of Photochemistry and Photobiology. B,* Biology 14(1-2): 3-22, 1992.

3. V. E. Tyler. What pharmacists should know about herbal remedies. *Journal of the American Pharmaceutical Association* NS36(1): 29-37, 1996.

4. Ergil K.V., Ergil, M., Furst, P.T., Gordon, N., Janzen, J.M., Sobo, E.J., and Sparrowe, L. *Ancient Healing: Unlocking the Mysteries of Health and Healing Through the Ages.* Lincolnwood, IL: Publications International, 1997.

5. S. Dev. Ancient-modern concordance in Ayurvedic plants: Some examples *Environmental Health Perspectives* 107(10): 783-789, 1999.

6. V. Badmaev, P. B. Kozlowski, G. B. Schuller-Levis, and H. M. Wisniewski. The therapeutic effect of an herbal formula Badmaev 28 (padma 28) on experimental allergic encephalomyelitis (EAE) in SJL/J mice. *Phytotherapy Research* 13(3): 218-221, 1999.

7. H. Drabaek, J. Mehlsen, H. Himmelstrup, and K. Winther. A botanical compound, Padma 28, increases walking distance in stable intermittent claudication. *Angiology* 44(11): 863-867, 1993.

8. J. A. Duke. *The Green Pharmacy: New Discoveries in Herbal Remedies for Common Diseases and Conditions from the World's Foremost Authority on Healing Herbs.* Emmaus, PA; New York: Rodale Press, 1997.

9. _____. *CRC Handbook of Medicinal Herbs.* Boca Raton, FL: CRC Press, 1985.

10. T. Johnson. *CRC Ethnobotany Desk Reference.* Boca Raton, FL: CRC Press, 1999.

11. D. E. Moerman. *Medicinal Plants of Native America, Research Reports in Ethnobotany; Contribution 2.* Ann Arbor: Museum of Anthropology, University of Michigan, 1986.

12. J. C. Aschoff. *Annotated Bibliography of Tibetan Medicine (1789-1995)* [Kommentierte Bibliographie Zur Tibetischen Medizin (1789-1995)]. Ulm, Germany; Dietikon, Switzerland: Fabri Verlag; Garuda Verlag, 1996.

13. World Health Organization. *WHO Monographs on Selected Medicinal Plants*. Geneva: World Health Organization, 1999.

Chapter 8

1. E. A. Gale and A. Clark. A drug on the market? *Lancet* 355(9197): 61-63, 2000.

2. J. I. Boullata and A. M. Nace. Safety issues with herbal medicine. *Pharmacotherapy* 20(3): 257-269, 2000.

3. A. K. Drew and S. P. Myers. Safety issues in herbal medicine: implications for the health professions. *Medical Journal of Australia* 166(10): 538-541, 1997.

4. Herbal Roulette. *Consumer Reports* 698-705, 1995.

5. D. V. C. Awang. Feverfew fever: A headache for the consumer. *HerbalGram* 43-46, 1993.

6. M. Blumenthal, W. R. Busse, and German Federal Institute for Drugs and Medical Devices. *The Complete German Commission E Monographs*. Austin, TX; Boston: American Botanical Council; Integrative Medicine Communications, 1998.

7. P. M. Wax. Elixirs, diluents, and the passage of the 1938 Federal Food, Drug and Cosmetic Act. *Annals of Internal Medicine* 122(6): 456-461, 1995.

8. B. Sibbald. New federal office will spend millions to regulate herbal remedies, vitamins. *CMAJ* 160(9): 1355-1357, 1999.

9. United States Pharmacopeial Convention. *The United States Pharmacopeia* (USP). Rockville, MD: United States Pharmacopeial Convention, 2000.

10. World Health Organization. *Guidelines for the Appropriate Use of Herbal Medicines, WHO Regional Publications, Western Pacific Series, No. 23*. Geneva: World Health Organization, 1998.

11. World Health Organization. *Quality Control Methods for Medicinal Plant Materials*. 115 Volumes. Geneva: World Health Organization, 1998.

Chapter 9

1. D. L. Sackett. *Evidence-Based Medicine: How to Practice and Teach EBM*. Edinburgh; New York: Churchill Livingstone, 2000, p. xiv, 261 pp.

2. OTA, Report of the Congress of the United States. "The Impact of Randomized Clinical Trials on Health Care Policy and Medical Practice." Washington, DC: U.S. Government Printing Office, 1983.

3. D. M. Eddy. Clinical decision making: From theory to practice. Anatomy of a decision. *Journal of the American Medical Association* 263(3): 441-443, 1990.

4. *The Review of Natural Products*. Vol. Oct. 1996-. St. Louis, MO: Facts and Comparisons, 1996.

Chapter 10

1. Monitoring W. C. C. f. I. D. *Global Intelligence Network for Benefits and Risks in Medicinal Products: Definitions.* World Health Organization (WHO) [cited January 31, 2001].

2. M. J. Cupp. Herbal remedies: Adverse effects and drug interactions. *American Family Physician. American Family Physician* 59(5): 1239-1245, 1999.

3. "Patients Being Treated for HIV Should Avoid St. John's Wort: NIH [News]." *Journal of the American Dental Association* 131: 439, 442, 2000.

4. J. Lazarou, B. H. Pomeranz, and P. N. Corey. Incidence of adverse drug reactions in hospitalized patients: A meta-analysis of prospective studies. *Journal of the American Medical Association* 279(15): 1200-1205, 1998.

5. J. M. Murphy. Preoperative considerations with herbal medicines. *Aorn Journal* 69(1): 173-175, 177-178, 180-183, 1999.

Chapter 11

1. D. Hoffmann. *The Herbal Handbook: A User's Guide to Medical Herbalism.* Rochester, VT: Healing Arts Press, 1988.

2. S. Ramsay. FDA cracks down on laetrile online. *Lancet* 356(9243): 1011, 2000.

3. Federal Trade Commission. "'Operation Cure.all' Targets Internet Health Fraud." <http://www.ftc.gov/opa/1999/9906/opcureall.htm>.

4. N. T. Landis. FTC zeroes in on Internet health fraud. Federal Trade Commission. *American Journal of Health-System Pharmacy* 56(15): 1489, 1999.

5. Online Health Information Seekers Growing Twice As Fast As Online Population. Cyber Dialogue, 2000 [cited January 26, 2001]. Available from <http://www.cyberdialogue.com/>.

6. "FTC Investigates Fraud on the Internet." *Direct Marketing* 1999, p. 14.

7. N. D'Epiro and S. D. Benjamin. Operation cure all: Taking action against wellness fraud. *Patient Care* 33(18): 23, 1999.

Chapter 12

1. T. Ferguson. Online patient-helpers and physicians working together: A new partnership for high quality health care. *British Medical Journal* 321(7269): 1129-1132, 2000.

2. D. Hoffmann. *The Herbal Handbook: A User's Guide to Medical Herbalism.* Rochester, VT: Healing Arts Press, 1988.

Chapter 13

1. D. J. de Solla Price. *Science Since Babylon.* New Haven, CT: Yale University Press, 1975.

2. B. L. Humphreys and D. E. McCutcheon. Growth patterns in the National Library of Medicine's serials collection and in Index Medicus journals, 1966-1985. *Bulletin of the Medical Library Association* 82(1): 18-24, 1994.

3. D. L. Sackett, W. M. Rosenberg, J. A. Gray, R. B. Haynes, and W. S. Richardson. Evidence based medicine: What it is and what it isn't. *British Medical Journal* (Clinical Research Edition) 312(7023): 71-72, 1996.

4. D. Kelsey. "IBM Software to Automate Study of Genetic Code." *MicroTimes*, August 2000.

5. J. W. Spencer and J. J. Jacobs. *Complementary Alternative Medicine: An Evidence-Based Approach.* St. Louis, MO: Mosby, 1999.

6. J. B. B. Ezzo, A. J. Vickers, and K. Linde. Complementary medicine and the Cochrane Collaboration. *Journal of the American Medical Association*, 1628-1630, 1998.

7. K. Linde and C. D. Mulrow. St John's wort for depression. *Cochrane Database Syst Rev* 7(2): CD000448, 2000.

8. W. D. Loub, N. R. Farnsworth, D. D. Soejarto, and M. L. Quinn. NAPRALERT: Computer handling of natural product research data. *Journal of Chemical Information and Computer Sciences* 25(2): 99-103, 1985.

Chapter 14

1. E. Ernst and M. H. Pittler. Efficacy of ginger for nausea and vomiting: A systematic review of randomized clinical trials. *British Journal of Anaesthesia* 84(3): 367-371, 2000.

2. K. Meyer, J. Schwartz, D. Crater, and B. Keyes. *Zingiber officinale* (ginger) used to prevent 8-Mop associated nausea. *Dermatology Nursing* 7(4): 242-244, 1995.

3. B. R. Cassileth, E. J. Lusk, T. B. Strouse, and B. J. Bodenheimer. Contemporary unorthodox treatments in cancer medicine: A study of patients, treatments, and practitioners. *Annals of Internal Medicine* 101(1): 105-112, 1984.

4. R. de Visser, D. Ezzy, and M. Bartos. Alternative or complementary? Nonallopathic therapies for HIV/AIDS. *Alternative Therapies in Health and Medicine* 6(5): 44-52, 2000.

5. M. Blumenthal, W. R. Busse, and German Federal Institute for Drugs and Medical Devices. *The Complete German Commission E Monographs.* Austin, TX; Boston: American Botanical Council; Integrative Medicine Communications, 1998.

Chapter 15

1. Burton Goldberg Group. *Alternative Medicine: The Definitive Guide.* Puyallup, WA: Future Medicine Pub., 1993.

2. N. J. Nelson. Scents or nonsense: Aromatherapy's benefits still subject to debate. *Journal of the National Cancer Institute* 89(18): 1334-1336, 1997.

3. N. Perry, G. Court, N. Bidet, and J. Court. European herbs with cholinergic activities: Potential in dementia therapy. *International Journal of Geriatric Psychiatry* 11(12): 1063-1069, 1996.

4. F. Mantle. Bach flower remedies. *Complementary Therapies in Nursing and Midwifery* 3(5): 142-144, 1997.

Chapter 16

1   T. Ferguson. *British Medical Journal* 321: 1129-1132, 2000.

Glossary

aromatherapy: An ancient practice of using essential oils extracted from plants to enhance mental, physical, and emotional well-being.

alternative medicine: Commonly used by consumers to refer to a diverse group of therapies not used or taught in mainstream medical schools and hospitals. Properly used to refer to a broad collection of therapies that are usually promoted as being valid treatment options to those offered by conventional medicine. The word *alternative* suggests that either might be effective.

allopathy: A word coined by Hahnemann, the founder of homeopathy, to distinguish his system from the "other," the conventional medicine of the eighteenth century. Today, the term *allopathy* is used to refer to Western science-based medicine.

Ayurvedic medicine: An ancient system of health care originating on the Indian subcontinent. It involves a holistic approach to health and healing. Its complex medicinal preparations are usually based on plant products.

bibliographic database: A database (e.g., MEDLINE) that serves as an index to the literature in a particular discipline. It includes citations that describe and identify titles, dates, authors, and other parts of journal articles, books, or other documents.

botanical name: Also referred to the scientific or Latin name. Each plant is given a unique name based on an internationally accepted system for naming unique and distinct plants, whether natural or cultivated.

botanicals: Plants used for their medicinal properties. Often used as a synonym for herbs.

chat room: An Internet-based tool in which communication takes place in real time, requiring participants to be online at the same time.

Commission E: Germany's Commission E was established to evaluate herbal preparations and to prepare monographs on their safety and efficacy. The findings, which have been published as monographs, are considered to be the closest thing yet to an authoritative source of information for the health care practitioner.

common name: A common name can be in any language and is usually selected based on the appearance of the flower or plant. One plant can have several common names.

complementary and alternative medicine (CAM): An umbrella term frequently used to cover both complementary and alternative therapies.

complementary medicine: Often used as a synonym for alternative medicine. However, complementary suggests therapies used alongside mainstream medical treatments. The term is more widely used in Europe than in North America.

curanderismo: A traditional Chicano healing system blending Native American and Hispanic healing techniques involving herbs, massage, diet, and magic.

dietary supplement: A dietary supplement is a product intended to supplement the diet. The U.S. Dietary Supplement Health and Education Act (DSHEA) of 1994, however, uses a much broader definition that allows the inclusion of vitamins, minerals, amino acids, and herbs.

eclectic movement: An important nineteenth-century American medical movement that introduced a range of herbs into common use.

empirical: Based on observation and experimentation. Used when a treatment works in practice but the basis for its success is not known.

ethnobotany: The study of the use of indigenous plants by people of various cultures in different parts of the world.

evidence-based medicine (EBM): This involves finding, appraising, and using the most up-to-date research findings as the basis for clinical decisions. The "gold standard" in EMB is usually the randomized controlled clinical trial (RCCT).

folklore: The traditional beliefs, legends, customs, and popular superstitions and legends of a people.

functional foods: Foods enriched with herbs or other dietary supplements.

good manufacturing practice (GMP): A set of regulations to control manufacturing conditions and to ensure that drugs are safe, pure, and effective.

GRAS list: A category of food additives that over time have been generally recognized as safe (GRAS) for human consumption. Several herbs are included because of their use in liqueurs and as components of natural flavorings.

herb: Commonly used to refer to any plant or plant part valued for its medicinal, savory, or aromatic qualities.

herbal: A book containing the names and descriptions of herbs with their properties and uses.

herbal medicine: A term often used interchangeably with herbalism. It implies the use of herbs according to the practice of Western scientific medicine.

herbalism: The study and use of plant material as food and medicine for healing and the promotion of health.

herbalist: A person who specializes in the use of plant material for healing and the promotion of health. Different cultures ascribe slightly different meanings to the term. For example, in China, an herbalist is considered to be a physician.

herbology: A term sometimes encountered as a synonym for herbalism. Presumably used because the addition of the suffix "-ology" sounds more scientific.

holistic medicine: A form of medical treatment that emphasizes the need to deal with the whole person. This includes the physical, mental, emotional, social, and spiritual aspects of health.

integrative medicine: Integrative medicine combines the use of both conventional and nonconventional therapies, based on evidence of efficacy and safety. It is evidence-based medicine, blurring the distinction between conventional and nonconventional medicine. This term is preferred by many health professionals.

investigational new drug (IND): Following animal testing, a drug company or other sponsor files an IND with the Food and Drug Administration (FDA) for permission to conduct human clinical trials.

Kampo: Japanese traditional medicine including the use of herbs.

mailing list: A mailing list, often erroneously referred to as a "listserv," is an automated system that allows people to send e-mail to one address, whereupon their messages are copied and sent to all of the other list subscribers.

materia medica: A Latin term meaning "medical materials," now usually referring to the materials used in pharmaceuticals.

monograph: By convention, each entry for a drug is called a monograph.

newsgroups: Internet newsgroups function as electronic bulletin boards.

nutraceuticals: Foods that are claimed to have medicinal benefit.

patent medicines: Preparations that were once marketed as wonder medicines capable of curing a wide range of diseases and conditions, from the common cold to cancer.

peer review: A process whereby an article, abstract, or some other written document is reviewed and evaluated for accuracy and quality by experts in the same field.

pharmacognosy: A word derived from the Greek *pharmakon,* or drug, and *gnosis,* or knowledge. The study of the physical, chemical, biochemical, and biological properties of drugs of natural origin.

pharmacopoeia: Books that describe and list standards for the identity, strength, quality, purity, packaging, and labeling of drug products.

phytomedicine: From *phyton,* Greek for plant. A term primarily used in European countries to describe a recognized category of plant-derived therapeutic drug products.

scientific medicine: Medicine practiced according to the scientific method, characterized by observation, measurement, and experiment.

search engine: Software that automatically searches for items either within a delimited area of the Web, such as a single server, or across the entire Internet.

taxonomy: The science of describing, naming, and classifying organisms and discovering their evolutionary relationships.

traditional Chinese medicine (TCM): An ancient healing system that encompasses a range of techniques and materials, including popular approaches such as acupuncture, massage, and herbal preparations, as well as more esoteric areas such as moxibustion and qi gong.

traditional medicine: Systems of healing that are not based on the Western scientific approach and are often centered around cultural beliefs and practices handed down from one generation to another.

Unani: A healing system derived from Greco-Arabic medicine that is currently practiced in India and Pakistan. It is primarily a type of herbal medicine.

Web site map: A list or other type of visual representation of a Web site's contents.

wildcrafting: The gathering of plant material from the wild.

Index

Page numbers followed by the letter "f" indicate figures; those followed by the letter "t" indicate tables.

Haworth Medical Information Sources
Sandra Wood, MLS, MBA
Senior Editor

HEALTH CARE RESOURCES ON THE INTERNET: A GUIDE FOR LIBRARIANS AND HEALTH CARE CONSUMERS edited by M. Sandra Wood. (2000). "A practical guide and an essential research tool to the Internet's vast and varied resources for health care has arrived—and its voice is professional and accessible . . . This comprehensive work is an important reference tool that is readable and enjoyable." *Elizabeth (Betty) R. Warner, MSLS, AHIP, Coordinator of Information Literacy Programs, Academic Information Services and Research, Thomas Jefferson University, Philadelphia, Pennsylvania*

EATING POSITIVE: A NUTRITION GUIDE AND RECIPE BOOK FOR PEOPLE WITH HIV/AIDS by Jeffrey T. Huber and Kris Riddlesperger. (1998). "Four stars! . . . A much-needed book that could have a positive impact on the quality of life for persons with HIV/AIDS. . . . Many of the recipes are old favorites that have been enhanced for the person with HIV. . . . All people with nutritional problems may also find this book helpful. It is not reserved solely for the person with HIV/AIDS." *Doody Publishing, Inc.*

HIV/AIDS AND HIV/AIDS-RELATED TERMINOLOGY: A MEANS OF ORGANIZING THE BODY OF KNOWLEDGE by Jeffrey T. Huber and Mary L. Gillaspy. (1996). "Provides the needed standardized terminology to describe large HIV/AIDS collections. . . . A welcome book for any cataloger, indexer, or archivist who is faced with organizing a mass of information that is growing very rapidly. . . . highly recommended for all librarians with extensive collections." *Booklist: Reference Books Bulletin*

HIV/AIDS COMMUNITY INFORMATION SERVICES: EXPERIENCES IN SERVING BOTH AT-RISK AND HIV-INFECTED POPULATIONS by Jeffrey T. Huber. (1996). "Provides a well-organized introduction to HIV/AIDS information services that will be useful to those affected by HIV disease, health care practitioners, librarians, and other information professionals. Appropriate for all libraries and an excellent reference resource." *CHOICE*

USER EDUCATION IN HEALTH SCIENCES LIBRARIES: A READER edited by M. Sandra Wood. (1995). "A welcome addition to any health sciences library collection. A valuable tool for both academic and hospital librarians, as well as library school students interested in bibliographic instruction." *National Network*

CD-ROM IMPLEMENTATION AND NETWORKING IN HEALTH SCIENCES LIBRARIES edited by M. Sandra Wood. (1993). "Neatly compacts information about the history, selection, and management of CD-ROM technology in libraries. . . Librarians at all levels of CD-ROM implementation can benefit from the solutions and ideas presented." *Bulletin of the Medical Library Association*

HOW TO FIND INFORMATION ABOUT AIDS, SECOND EDITION edited by Jeffrey T. Huber. (1992). "Since organizations and sources in this field are constantly changing, this updated edition is welcome. . . . A valuable resource for health or medical and public library collections." *Booklist: Reference Books Bulletin*